Claiming Identity Through Redefined Teaching in Construction Programs

Sherif Mostafa
Griffith University, Australia

Payam Rahnamayiezekavat
Western Sydney University, Australia

A volume in the Advances in Educational Technologies and Instructional Design (AETID) Book Series

IGI Global
DISSEMINATOR OF KNOWLEDGE

Published in the United States of America by
IGI Global
Information Science Reference (an imprint of IGI Global)
701 E. Chocolate Avenue
Hershey PA, USA 17033
Tel: 717-533-8845
Fax: 717-533-8661
E-mail: cust@igi-global.com
Web site: http://www.igi-global.com

Library of Congress Cataloging-in-Publication Data

Names: Mostafa, Sherif, editor. | Rahnamayiezekavat, Payam, editor.
Title: Claiming identity through redefined teaching in construction programs / Sherif Mostafa and Payam Rahnamayiezekavat, Editors.
Description: Hershey PA : Information Science Reference, [2020]
Identifiers: LCCN 2018054451| ISBN 9781522584520 (hardcover) | ISBN 9781522584537 (ebook) | ISBN 9781522585077 (softcover)
Subjects: LCSH: Construction industry--Management--Study and teaching.
Classification: LCC TH438 .C575 2019 | DDC 624.071--dc23 LC record available at https://lccn.loc.gov/2018054451

This book is published in the IGI Global book series Advances in Educational Technologies and Instructional Design (AETID) (ISSN: 2326-8905; eISSN: 2326-8913)

British Cataloguing in Publication Data
A Cataloguing in Publication record for this book is available from the British Library.

Advances in Educational Technologies and Instructional Design (AETID) Book Series

ISSN:2326-8905
EISSN:2326-8913

Editor-in-Chief: Lawrence A. Tomei, Robert Morris University, USA

MISSION

Education has undergone, and continues to undergo, immense changes in the way it is enacted and distributed to both child and adult learners. In modern education, the traditional classroom learning experience has evolved to include technological resources and to provide online classroom opportunities to students of all ages regardless of their geographical locations. From distance education, Massive-Open-Online-Courses (MOOCs), and electronic tablets in the classroom, technology is now an integral part of learning and is also affecting the way educators communicate information to students.

The **Advances in Educational Technologies & Instructional Design (AETID) Book Series** explores new research and theories for facilitating learning and improving educational performance utilizing technological processes and resources. The series examines technologies that can be integrated into K-12 classrooms to improve skills and learning abilities in all subjects including STEM education and language learning. Additionally, it studies the emergence of fully online classrooms for young and adult learners alike, and the communication and accountability challenges that can arise. Trending topics that are covered include adaptive learning, game-based learning, virtual school environments, and social media effects. School administrators, educators, academicians, researchers, and students will find this series to be an excellent resource for the effective design and implementation of learning technologies in their classes.

COVERAGE

- Adaptive Learning
- Web 2.0 and Education
- Classroom Response Systems
- Bring-Your-Own-Device
- Collaboration Tools
- Virtual School Environments
- Higher Education Technologies
- Curriculum Development
- Digital Divide in Education
- Instructional Design

IGI Global is currently accepting manuscripts for publication within this series. To submit a proposal for a volume in this series, please contact our Acquisition Editors at Acquisitions@igi-global.com or visit: http://www.igi-global.com/publish/.

The Advances in Educational Technologies and Instructional Design (AETID) Book Series (ISSN 2326-8905) is published by IGI Global, 701 E. Chocolate Avenue, Hershey, PA 17033-1240, USA, www.igi-global.com. This series is composed of titles available for purchase individually; each title is edited to be contextually exclusive from any other title within the series. For pricing and ordering information please visit http://www.igi-global.com/book-series/advances-educational-technologies-instructional-design/73678. Postmaster: Send all address changes to above address.

Titles in this Series

For a list of additional titles in this series, please visit:
[go here and find specific BS URL: http://www.igi-global.com/book-series/]

Handbook of Research on Diverse Teaching Strategies for the Technology-Rich Classroom
Lawrence A. Tomei (Robert Morris University, USA) and David D. Carbonara (Duquesne University, USA)
Information Science Reference • copyright 2020 • 450pp • H/C (ISBN: 9781799802389) • US $265.00 (our price)

Emerging Technologies in Virtual Learning Environments
Kim Becnel (Appalachian State University, USA)
Information Science Reference • copyright 2019 • 348pp • H/C (ISBN: 9781522579878) • US $205.00 (our price)

English Language Teaching in a Post-Method Paradigm
Paulette Joyce Feraria (The University of the West Indies, Jamaica)
Information Science Reference • copyright 2019 • 426pp • H/C (ISBN: 9781522592280) • US $175.00 (our price)

Cases on Digital Learning and Teaching Transformations in Higher Education
Rebecca J. Blankenship (Florida Agricultural and Mechanical University, USA) and Charlotte Baker (Virginia Polytechnic Institute and State University, USA)
Information Science Reference • copyright 2019 • 266pp • H/C (ISBN: 9781522593317) • US $175.00 (our price)

Handbook of Research on Innovative Digital Practices to Engage Learners
Prince Hycy Bull (Gardner-Webb University, USA) and Jared Keengwe (University of North Dakota, USA)
Information Science Reference • copyright 2019 • 468pp • H/C (ISBN: 9781522594383) • US $275.00 (our price)

For an entire list of titles in this series, please visit:
[go here and find specific BS URL: http://www.igi-global.com/book-series/]

IGI Global
DISSEMINATOR OF KNOWLEDGE

701 East Chocolate Avenue, Hershey, PA 17033, USA
Tel: 717-533-8845 x100 • Fax: 717-533-8661
E-Mail: cust@igi-global.com • www.igi-global.com

Table of Contents

Detailed Table of Contents

This chapter outlines culminates experience and thinking from two environments: 1) the construction contracting industry and 2) the built environment classroom – undergraduate and graduate. The authors' combined teaching experience covers more than two decades, but they do not pretend their knowledge is exhaustive. Others know more. Their desire is to add to their thoughts. They first address what the contractor desires of construction graduates. From there, they proceed to list focuses they believe important in sequential order for the teaching and learning process: 1) vocabulary, 2) conceptual frameworks, 3) iterative problem solving, 4) writing. After that, they describe their preferred delivery method: formative teaching, learning, and assessment.

Although group work has been proven to be an effective method for enhancing active learning in the higher education, optimum planning is crucial for successful implementation. A deep understanding of teamwork dynamics and creation of inclusive environments helps groups to demonstrate their optimum performance and output. On this basis, the current research focuses on the important challenge of gender inclusiveness and required teacher interventions to encourage that. Towards this aim, three research hypotheses are developed and tested using student performance data in a series of individual, group, and hybrid assessment. Findings show the significantly different performance of female and male students in group activities. It is also found that instructor interventions to form gender-inclusive groups significantly improve group performance and output. This works contributes to the higher education literature by exploring dynamics of collaborative learning and interfaces with gender inclusiveness. Educators can utilize the findings to better design and implement team activities.

Education of the built environment is moving towards more collaborative practices. The intent behind the collaborative approach of teaching is to encourage students to explore the unknowns and unravel the problems themselves with the professor acting as the facilitator. This chapter presents the case for collaborative pedagogical approaches that was adopted to teach students from two geographically distant universities. The occasion used was preparation of student teams for Associated Schools of Construction (ASC) student competitions. The underlying philosophy behind this initiative is to provide the context for the future leaders in the AEC to learn by doing in a safer and potentially rewarding environment.

Students often feel that assessments are used primarily for grading purposes only. There is anecdotal evidence too that students are being over assessed in UK Higher Education (HE) given that it is common place for the curricula to be designed with formative and summative assessments in mind. This reflective case study aims to establish the link between the attainment of independent learning outcome and a new form of assessment that is based on three years of teaching a Project Management unit in two internationally leading Schools of Mechanical, Aerospace, and Civil Engineering in England and Singapore. Data from the observation of an operational unit specification reveals that depending on the timing within the schedule of the unit, summative assessment with appropriate feedback can be re-purposed as formative and summative. The implication being a paradigm shift in the design of unit specification for the purpose of attaining Intended Learning Outcomes (ILO) that enhance deep learning.

Despite a growing awareness of climate change, there is little evidence that this is being addressed in cities and built environments. Events such as flooding in Houston, USA; landslides in Free Town, Sierra Leone; and water shortages in La Paz, Bolivia and Cape Town in South Africa demonstrate that it is increasingly important that climate change is understood and addressed in built environments to ensure that they become more resilient. This chapter introduces climate change and outlines the implications of this for built environments. It describes measures that can be incorporated into built environments to enable them to adapt to projected climate changes. Understanding climate change and preparing for this by developing built environments that are more resilient will be an increasingly valuable and important skill. Reading this chapter will support the development and refinement of skills and knowledge in this area and it is an essential reference for built environment students and practitioners.

This chapter is aimed at appraising the reader with the course delivery methodology/ teaching pedagogy practiced at School of Construction at RICS School of Built Environment, Amity University. The chapter highlights the innovative tools and techniques devised by the school for enhancing a students' learning process and ingraining the thoughts for sustainability in construction at nascent stages of construction management. The course is designed to provide the student with an understanding of sustainable construction principles and communicating current sustainability practice to various stakeholder groups.

The construction industry is currently struggling due to the ever-increasing complexity of a building and the highly fragmented nature of the construction industry. In response to the current problems, building information modelling (BIM) has been adopted to improve productivity and deliver a quality building to a client. The academia also has been striving to embed BIM education into the exiting curriculum to accommodate the needs of the construction industry. However, BIM is currently taught as a simple design tool rather than a methodology to improve productivity in construction projects. Although there have been various attempts to integrate BIM courses into the existing curriculum, a lack of research has been conducted regarding how BIM adoption and implementation strategy should be utilized for a BIM integrated curriculum. Thus, this chapter will provide insights for updating and developing BIM courses.

The chapter summarizes the educational pedagogy researched and developed in the OnSite project, a multi-year trial of a blended virtual learning environment, situating it in the context of immersive learning environments generally and discussing the specific challenges in designing and creating environments suitable for introductory construction courses. It documents and reports the challenges in creating a virtual learning environment (VLE) for use within an introductory construction course for architecture and building students at the University of South Australia. In addition, the chapter will reflect on issues of technical development for immersive learning environments, discussing the purpose and value of high-fidelity modelling, texturing, and lighting to achieve learning "authenticity." Finally, the chapter looks at the implications for VLEs of this sort in terms of larger issues regarding the potential for game engine (GE) environments as cooperative spaces within the AEC industry.

The chapter examines the use of visual tools in inquiry-based learning in engineering education. Engineering information has traditionally been communicated using engineering drawings. The advancement of technology has empowered engineers to work with data to generate graphics where visual tools have been successfully employed in the industry. However, many educational institutions still rely on text-based learning, which rely on linguistic abilities to convey complex concepts. Using program logic theory, the authors evaluate how a poster can help learners acquire knowledge and know-how to represent complex ideas in a logical and consistent manner. Students reported benefits in poster exercise compared with a written assignment. This chapter also discusses the educational and managerial significance of using visual tools to greater effectiveness and the importance of communicating technical information.

Preparing project management students for their future life and work requires actively developing and evidencing a suite of transferable skills and attributes. This chapter reports on how a student-centered pedagogy, which included the use of guided sequential exercises, and the collection of instantaneous student responses through a personal response system (PRS), was implemented in a large first year undergraduate Project Management course. The students' perceptions of this pedagogy demonstrate that they found the pedagogical approach supported their learning and fostered deeper engagement in the course, with the most useful aspect of the course perceived to be its interactive nature. The chapter affirms the importance of giving life to a learning orientation conception of learning. The chapter has implications for ensuring learner engagement in the particular discipline of project management and for good practice in large class context in higher education.

Foreword

Peak professional bodies in the Construction industry such as the Australian Institute of Building (AIB) and the Australasian Universities Building Education Association (AUBEA) have long advocated the value of specialised education for Building and Construction Management. In line with this objective, this book is an endeavour to firstly promote Construction Management as a specialised area of education and secondly share the emerging methods crafted within the Construction Management education environment to serve its peculiar training traits. I re-iterate AIB's motto of 'Advance through Learning' to emphasise the need for documenting best teaching practices that are designed to more effectively deliver tertiary building courses. It was interesting to see that on top of coverage of modern trends in Construction Education the book touch bases delicate matter of gender inclusiveness. A few keywords that come to mind when identity of Construction industry is claimed to form through continuous improvement of our approach to teaching – I interpret 'redefined' in the title of the book to amount in everlasting cycles of rethinking how to produce better building professionals – are Building Information Modelling (BIM), sustainability, urban resilience and a range of project management skills. To no surprise, there are Chapters in the book that present details of techniques that have been designed and implemented towards enhancing learning on topics such as BIM, sustainable built environment, building in extreme climates, and data visualisation for project management. Furthermore, the book reflects on application of several state-of-art teaching methods to deliver Construction-specific knowledge and skills. It is of great importance to the leadership of tertiary education, as demonstrated in different Chapters, that the transition to advanced teaching in Construction Management programs such as utilising the version of flipped classroom adapted for Construction is resource intensive. Academics as the facilitators of active learning are key resources essential for successful adoption of new learning and teaching paradigms. Investment in transforming construction tertiary teaching is by no means a one-off event and requires continuous allocation of resources that will eventually return to our industry in form of competent graduates ready to tackle challenges of 21st century construction.

Josua Pienaar
Central Queensland University, Australia & Australasian Universities Building Education Association, Australia

Preface

Construction projects are notorious for their one-of-kindness that requires customised management strategies to keep ongoing field activities in order. Nevertheless, unique attributes of practice are manifested in the curriculum educators develop for Construction programs. A collage of management, engineering and technology, material sciences, commerce, economics, accounting, law, together with construction-specific topics such quantity and building surveying hinder deployment of a "one-size-fits-all" approach to the teaching of Construction Management. This challenge is exclusive to Construction Management programs as other disciplines have consolidated a mainstream teaching methodology to cater for their specific pedagogical needs. Law and debates, Architecture and design studios, Medicine and hospital rounds, Engineering and laboratory demonstrations, Business and simulation facilities are logical combinations coupling a teaching methodology with the subject it suits the best. Accordingly, this publication endeavours to highlight the need for recognising most effective approaches towards the delivery of diversified knowledge areas incorporated in Construction Management programs across the globe. We argue that there should be a "signature" teaching method, at least in each knowledge area, to shape the identity of Construction Management as an independent discipline. The outcome is a degree of teaching consistency among many components of typical Construction Management programs that assures students navigate smoothly from one unit to another. The enabling step towards a harmonised teaching, with no doubt, is a collection of successful cases to provide a basis for comparison and then synthesis of a standard method.

We started this book project with one explicit target in mind: Investigation of Construction Management's unique learning and teaching ecosystem. We aimed to promote the idea that while our curriculum overlaps with many other disciplines, it adopts a particular teaching paradigm. To demonstrate this, we invited Construction Management educators to elaborate on the strategy they adopt in their teaching. We set a simple criterion for accepting Chapters for publication in this book. The only requirement to suit the book was for the Chapter to reflect difficulties that the authors have encountered and the solutions they deployed for the development of Construction Management pedagogy.

The book is meant to turn into a reliable stepping stone for early career academics to grasp ideas about the design and delivery of newly emerged courses assigned to them. The book would also be a valuable self-assessment tool for academics delivering a course in areas similar to those covered in the book. Furthermore, for the purpose of benchmarking, the book is of interest to the leadership team in Schools that offer Construction programs as well as accrediting bodies. Finally, students can consult the book as a rich learning source that presents state-of-the-art pedagogy covering the critical areas of the contemporary practice of Construction Management. In summary, the audiences of this book include educators, curriculum developers, learning technology designers and researchers in the construction management space. These audiences can use this book content to improve their understanding of the teaching and learning strategies suitable for construction management units.

ENGAGING STUDENTS' LEARNING IN THE BUILT ENVIRONMENT THROUGH ACTIVE LEARNING

Scott depicts a clear image of what student centred learning looks like in the realm of Construction Management. He, as repeatedly mentioned in this book, pinpoints the changes in the construction practices, and consequently teaching approaches, to avoid deductive doctrine where students are the repository of theories and applications are marginalised. He postulates that an inductive approach should inform teaching in the Built Environment where students are firstly exposed to construction tasks and associated challenges and only then explore theories that they need to manage the tasks and resolve the issues. This Chapter confronts Construction educators with practical challenges to actively engage students in the process of collaborative learning. Scott covers significant challenges such as transforming classroom to social space and balancing learning workload between class group learning and out-of-class activities and assignments.

A CONSTRUCTION MANAGEMENT EDUCATION FOCUS AND PROCESS DIRECTION

This chapter by Stevens and Day contradicts attributes of successful contractors with how they are cultivated in students in classroom settings. The Chapter provides a useful summary of desirable knowledge sought by contractors as employers of our graduates. The authors compile the list by reflecting on their extensive industry experience. Then they explain their four-step model to teach students to acquire the desired knowledge. The consecutive steps comprising their model include 1) Vocabulary, 2) Conceptual Frameworks 3) Iterative Problem Solving, and 4) Writing.

The chapter concludes by elaborating on 'formative process', argued by the authors to be the element that binds the four components of the model and transforms ordinary students into work-ready graduates.

This chapter signposts the dynamics of the construction industry and the corresponding duty of the academics to adapt their teaching with the new realities. They indicate that despite many uncertainties in our industry, it is certain that the future will not follow the past. This emphasises the need for fostering self-resilient students, and we recommend that the proposed model is a solid starting point to achieve this objective.

GROUP DYNAMICS IN HIGHER EDUCATION: IMPACTS OF GENDER INCLUSIVENESS AND SELECTION INTERVENTIONS ON COLLABORATIVE LEARNING

The construction industry is notorious for its male-dominant environment. The study by Arashpour, Lamborn and Farzanehfar resonates with this trait of our industry and indirectly justifies the ongoing efforts towards equity and gender balance. The presented study scientifically analyses the dynamics of group work among undergraduate students. The findings suggest that instructor interventions, in the form of enforcing gender inclusiveness, will result in improved performance of the group for male and female students. This observation is analogous to policy interventions to encourage gender balance in construction projects.

This chapter sheds light on the impact of gender inclusiveness and successful execution of collaborative learning strategies. In other word, Arashpour, Lamborn and Farzanehfar remind educators that other than properly designed collaborative learning activities, overlook of factors such as gender inclusiveness may result in suboptimal learning outcomes. The Chapter also spells out the central role of educators in the delivery of collaborative learning. Indeed, the transition to modern teaching practices such as collaborative learning consolidates the role of educators rather than fading their engagement in the process of learning. The implication for universities is that modern teaching techniques are more resource intensive in comparison to traditional teaching styles.

LEARNING THROUGH SUCCESSFUL DIGITAL OPPORTUNITIES FOR EFFECTIVE COMPETITION PREPARATIONS

Scott and Shelbourn explain in this chapter how collaborative pedagogy equips students to delve into unknowns and work out required solutions. They examined different aspects of collaborative pedagogy in preparing student team for an inter-school contest and established that their approach provided a safe and rewarding environment for nurturing leaders of the future. One pivotal implication for educators interested in deploying collaborative pedagogical approach was their re-defined role as facilitators. Successful fulfilment of this different role requires educators to develop new skills if collaborative pedagogy is intended to be used. The authors finish the chapter with some practical recommendations for execution of the collaborative pedagogical approach. Teambuilding is the essence of the advice given with particular emphasis on allowing a loose structure for participating teams.

RE-PURPOSING SUMMATIVE ASSESSMENT AS FORMATIVE: A REFLECTIVE GUIDE TO FACILITATING DEEP LEARNING

Ejohwomu touches the delicate issue of over-assessment to the extent that summative assessment refuses to serve its purpose and operates in detriment of effective learning. The guideline the book author presented enables academics to revitalise the portion of their assessment regime that falls within the category of over-assessment, in the form of formative assessment. Ejohwomu advises that time is of the essence for the repurposing of summative assessment. Another key consideration for shifting to formative assessment is the class size. This practical reflection by the author is of great value to the readership of this book knowing that the three yeas journey reflected on by the author has secured him a fellowship with the Higher Education Academy of the UK.

CLIMATE CHANGE IN THE BUILT ENVIRONMENT: ADDRESSING FUTURE CLIMATES IN BUILDINGS

The chapter is relying on the catastrophic aftermath of several recent natural disasters across the world. Gibberd argues that despite raising public awareness about climate change, the current practice of building does not consolidate resilience against negative consequences of unprecedented extreme climate events. Accordingly, Gibberd discusses the verity of skills and knowledge that are essential to be taught to the built environment students to leverage resilience in their future practices.

SUSTAINABLE PRACTICES FOR THE BUILT ENVIRONMENT

Smart buildings and cities are familiar terminologies for academics who are tasked with teaching sustainability in many different academic programs. Thanikal, Bhatnagar,

Jain and George elaborate in this chapter how the curriculum across a typical School of Construction can accommodate sustainability as an integral component like other traditionally recognised areas such as cost, time and quality. Given the rapid urbanisation of India and, as indicated by the authors, the lack of provisions for the protection of the natural environment in the Government policies and regulations, Indian universities are to shoulder higher responsibilities to prepare next generation of the built environment professionals to address the overlook of the environment in India's urban development. The detailed description of the strategy deployed by RICS School of Built Environment to instil sustainability in the curriculum of an MBA program specialling in Construction Project Management is a unique endeavour that could inform similar initiatives to develop a postgraduate Construction degree.

INTEGRATED BIM EDUCATION IN CONSTRUCTION PROJECT MANAGEMENT PROGRAM

Kim, Mostafa and Park point to the emergence of BIM and its uptake by the industry to handle complex building designs and broken communications among stakeholders throughout the project life cycle. They correctly argue that the industry's enthusiasm for BIM urges BIM-embedded curriculum. They use well-known 'BIM adoption' terminology, expectedly prevalent in industry communiques, to discuss their strategies for developing BIM courses. This chapter outlines a distributed approach, in contrast to a centralised tactic, where different subjects were examined for their readiness to accommodate BIM related modules. The chapter reports students' satisfaction with their BIM experience and highlights the inevitable correlation between teaching BIM and utilising real-world projects as part of students' learning journey.

ONSITE: THE VIRTUAL SITE VISIT AS AN ENVIRONMENT FOR CONSTRUCTION LEARNING

This chapter incorporates work of Pickersgill, Rameezdeen and Harvey on development of a virtual learning environment to enrich the delivery of an introductory Construction course. The chapter will cover the development of a curriculum that accommodates virtual reality as an integral part of students' learning as well as technical challenges one should expect in developing the virtual environment itself. *OnSite* is a good example of how virtual reality could be used to teach construction specific topics such as different pieces of Standards that are required to be complied with in the real world. The value of virtually accessing a site is clearly elaborated by comparing against the resource intensive alternative of physical site visits. It is of great importance that the authors attribute the varying success of *OnSite* trials in different universities to the way the tool was integrated with the associated course. Indeed, success was guaranteed when the curriculum was developed around features of *Onsite*. On the other hand, endeavours to use *OnSite* as an addition to already existing learning materials marked lower success rate. This observation is in the same vein with the perspective implied in 'Learning through Successful Digital Opportunities for Effective Competition Preparations' stating that construction educators have to continuously improve their teaching competencies to maintain their ability to upgrade their teaching with same pace changes happens in the industry.

VISUAL TOOLS FOR ANALYSING AND ORGANISING ENGINEERING DATA AND INFORMATION

Modern, alternative means of communicating project information is the central tenet of this Chapter authored by Hui, Zarei, and Duffield. The emphasis is on the shift from textual to visual representation of data and a case study of an endeavour to make students rethink traditional project communication platforms and explore the potential to master visualisation as a common language to facilitate communication between project stakeholders. In particular, the chapter presents data to support the claim that the use of visual tools to aid teaching project management proved to enhance students learning. Interestingly, the task to extract, digest and present a complex project's information essential for systems management and procurement strategy in poster format, not only improved students' learning but also boosted their confidence in dealing with others.

STRATEGIES FOR EFFECTIVE TEACHING IN PROJECT MANAGEMENT

Gharaie and Wingrove stress the importance of student centred pedagogy and share their insight on how technology may enable educators to effectively interact with students in large classrooms. Their experiment with the use of Personal Response System in a Project management course can readily be extended to other Construction subjects as long as their approach for the development of a series of class exercises is precisely replicated. The authors demonstrate in the chapter, for example, how simple adjustments in the instruction of the class exercise on Critical Path Method prepared the environment for better student engagement using the polling system. Aligned with other chapters, this chapter is another testament to the importance of modernising Construction teaching style in order to prepare professionals capable of handling future unknowns. Gharaie and Wingrove advise that students themselves expect the curriculum to bridge between learning outcomes and their future profession.

This brief overview of the contents demonstrated how this book provides a platform to publicise and hence increase widespread adoption of innovative learning and teaching mechanisms developed at different institutions to address the ever-changing attributes of work-ready graduates in the field of Construction Management. The emphasis of the book is to demonstrate 'signature' learning and teaching strategies specific to the construction curriculum that collectively pictures the independent identity of construction academic programs otherwise would have been perceived as a revamp of several other established disciplines. The construction industry is reforming itself to embrace new norms such as social responsibility, deploy modern construction methods such as modular building, and modernise/smart construction contracts. Accordingly, universities as the training powerhouse are to reform their teaching paradigm at both the program and course level. This results in a need to share successful approaches examined in various domains comprising Construction body of knowledge and the relevant skillset.

Sherif Mostafa
Griffith University, Australia

Payam Rahnamayiezekavat
Western Sydney University, Australia

Acknowledgment

This book has been developed from research conducted by researchers from 11 different universities. We want to thank the book chapters authors from the Aston University, University of South Australia, Griffith University, Western Sydney University, the University of Melbourne, RMIT, Monash University, Technological University Dublin, University of Salford, The University of Manchester, and Amity University. We also extend our gratitude and acknowledgement to the reviewers for their time and constructive feedback provided on the chapters' submissions. Special thanks to Professor Josua Pienaar for providing an outstanding foreword.

We are sincerely thankful to the IGI Development Team for their assistance throughout the book development project. Lastly, we wish to express our thanks to the School of Engineering and Built Environment, Griffith University and the School of Computing, Engineering and Mathematics, Western Sydney University for the support provided throughout the book concept until its publication.

Chapter 1
Engaging Students' Learning in the Built Environment Through Active Learning

Lloyd Martin Scott
https://orcid.org/0000-0001-6305-3749
Technological University Dublin, Ireland

ABSTRACT

This chapter addresses the position of learning, teaching, and assessment in education with the particular emphasis on higher education (HE) in the built environment (BE) but also embeds the context of the contemporary approaches that have emerged in the BE which are built on a solid educational underpinning. The conceptions of "active learning" are addressed from the perspective of what the literature refers to but also some significant reference to action research adopted, rolled out, and evaluated in undergraduate built environment education by the author. The "learning by doing" mantra of among the BE educational community has begun to make inroads. There is a more engaging approach from academics to support learners. An identification of the areas where improvement may be achieved into the future and the possible areas where research might be explored to address and solve some of these pertinent issues.

INTRODUCTION

I know I cannot teach anyone anything. I can only provide an environment in which we can learn -Carl Rogers

DOI: 10.4018/978-1-5225-8452-0.ch001

This chapter builds on the personal, educational and cultural contexts of the author. It addresses the position of learning, teaching and assessment in education with the particular emphasis on Higher Education (HE) in the Built Environment (BE) but also embeds the context of the contemporary approaches that have emerged in the BE which are built on a solid educational underpinning. The conceptions of 'active learning' are addressed from the perspective of what the literature refers to but also some significant reference to action research adopted, rolled out and evaluated in undergraduate built environment education by the author.

Since the early 1980's numerous educational leaders in the field of HE made recommendations and urged academics to adopt more student centred learning approaches, including the adoption of more active learning methodologies. Bringing to the classroom ways to more actively involve and engage students in the process of learning. The 'learning by doing' mantra of among the BE educational community has begun to make inroads. There is a more engaging approach from academics to support learners. Although we consider the concept of student centered learning (SCL) to be a relatively new one, particularly in HE, interest in it has been long-standing among educators in all realms of education. In looking to characterise the various ways that SCL is defined, and how the learner experience might be enhanced through the use of methodologies that improve learning it is necessary to consider how academics approach learning and teaching. To contextualise student centered approaches, reflecting on some of the developments that have taken place in HE education is necessary. In particular, a focus on the BE, and how the experience of learners has been enhanced, through a variety of differing approaches to learning, teaching and assessment. An identification of the areas where improvement may be achieved into the future and the possible areas where research might be explored to address and solve some of these pertinent issues.

It looks at the pedagogic approach necessary for teacher/lecturers to consider, in the modern era of change in HE, in order that they can engage in practices that will foster and encourage deep and meaningful learning by students. The role of technology is briefly reflected upon and the contribution it can make to the learner experience identified. In particular, assessment for learning is investigated and defined, along with exploring teaching excellence as a concept to encourage and engaged academics in best practice through assessment as a theme while acknowledging the need to explore reflective practice in approaches by teachers/lecturers. To begin a speculative look back at the 12th Century HE educational experience of students and what they might have encountered back then is a good starting point. This is followed by a brief comparison with what one might expect to find in our modern 21st century educational establishments.

Assessment, which is considered an active learning methodology, has always been considered an important part of the educational process but it has rarely been discussed in the discipline or subject context with the exception of some of the more practical aspects. The work of Wright and Jones (2007) where case study examples of good practice in the BE are offered appears to be one of the few discipline resources that have been published in the BE area. The discipline research literature on the topic has been very thin on the ground although there is agreement that this is a very important aspect of the work of academics (Peel, 2010). Within the BE in Ireland there is very little research on assessment with respect to undergraduate programmes. A small amount of research has been produced with respect to postgraduate assessment and the work of the author addresses the judgmental aspects of assessment in respect of taught programmes and the embedding of formative assessment led learning strategies (FALLS) (Scott & Fortune, 2009, 2011 and 2012).

BACKGROUND

Traditional BE instruction has been adopted with a deductive approach, more often beginning with theories and progressing to the applications of those theories. Alternative learning and teaching approaches are more inductive. Topics are introduced by presenting specific observations, case studies or problems, and theories are taught or the students are helped to discover them only after the need to know them has been established. This chapter refers to several of the most commonly used inductive teaching methods, including inquiry learning, problem-based learning, project-based learning, case-based teaching, discovery learning, and just-in-time teaching. The chapter offer a definition of each method, highlights commonalities and specific differences, and reviews research on the effectiveness of the methods. As the focus of this chapter is on active learning the underpinning principles that support this pedagogical approach are emphasised.

While the strength of the evidence varies from one method to another, inductive methods are consistently found to be at least equal to, and in general more effective than, traditional deductive methods for achieving a broad range of learning outcomes.

What is Active Learning and Why Should We Adopt It?

IT is learning by doing where the learner is placed at the centre of the learning. Research shows that active learning is much better recalled, enjoyed and understood. Active methods require us to 'make our own meaning', that is, develop our own conceptualisations of what we are learning. During this process we physically make neural connections in our brain, the process we call learning. Passive methods such as listening do not require us to make these neutral connections or conceptualisations. Active methods also:

- Give the learner feedback on their incomplete understandings and encourage them fix this, for example by helping each other.
- Give the teacher feedback on which learners understand, and who needs help
- Develop thinking skills such as analysis problem solving, and evaluation
- Help learners to use their learning in realistic and useful ways, and see its importance and relevance
- Can be more fun!

Good students may create meaning from passive methods, but weaker students do not but more often need to grasp a principle by way of 'doing'. Both types of student improve their learning enormously when they are required to use and apply it. When active methods are compared with traditional methods in careful control group and experimental group comparisons, active methods produce much better achievement. This is true even though the control group (traditional didactic methods) has the same teaching time as the experimental (active learning) group. This is also true at every academic level.

So the sceptic will say 'well if Active Learning works, why don't more lecturers use it?'

So the research tells us that we tend to teach the way we were taught ourselves, rather than in the way that works best or is most appropriate for the context. We as knowledge experts know too much, and rather enjoy explaining. So when you set activities, listen carefully to learners as they work, this can be even more enjoyable and less hard work than explaining, and the feedback is very informative.

THE CHANGING LANDSCAPE OF HIGHER EDUCATION

In setting the scene for section of this chapter, a trip back in time, back to the 12th Century in fact, and a wander down the corridors of the University of Bologna (Universites Bonoiensis), regarded as the first university in the Western world, is appropriate. On looking into one of the lecture halls you would probably see a teacher/lecturer talking at a room full of gown clad bored-looking students. He might be tediously speaking about particular subject matter that the student body may not fully understand or even care about. Perhaps, he may ask the occasional question, and without response almost certainly provide the answer himself. Some students may be taking notes, some verging on falling asleep and others most certainly staring vacantly into space. The teacher/lecturer might close the proceedings with a slight recap followed by readings assigned for the next lecture.

HE today does not completely follow the above model and, in particular, elements in Built Environment education across the globe do not. Lecture halls today are more appropriately designed with tiered seating, the comfort factor inbuilt, along with environmentally friendly whiteboards. In many instances the inclusion of up to date technology to offer assistance to the teacher/lecturer is very much in evidence, from the data projector to WIFI access for laptops. In some instances, we have specially designed 'learning spaces' that are adaptable and flexible to accommodate all perceived needs. The student body dresses differently; gone are the gowns, replaced by fashion conscious students with mobile technologies attached. Technologically equipped, as many of our lecture halls are today, a continued stroll down the corridors of our 21st Century HE institutions provides a good chance of meeting the descendants of our Bolognesi ancestors staring vacantly into space as the traditional instruction paradigm lecturer/ teacher, stuck in a time warp, drones on about some particular built environment related principle or concept. However, one might also experience learners engaged in discussing a problem in the corridors or even some learners surrounding a laptop sourcing the recent iteration of resources uploaded to the Learning Management System for a particular subject area.

We might well ask – so why make any changes? What is the response to the traditional teacher/lecturer who defends the position of tradition, 'this is how I was taught and look at how I turned out' or 'it has always been done this way and if it is not broken then why change'. There are many responses one might make to our traditionalist colleague, enough to write a thesis on it own, however, while excellent graduates have taken their place in society, the question must be asked, has this happened because of their traditional education experience or in spite of it? Also, unlike the teacher/lecturer of the 12th Century, today much more is known about how people learn and the educational practices that facilitate learning, including student centered learning and more experiential learning methodologies. Finally, it must be considered, is the traditionalist stuck in a time warp and simply uneasy about making change? It is recognised that in HE some academics teach students without having much formal knowledge of how students learn (Fry *et al.*, 1999, others).

Brown *et al* refer to the complex nature of learning and define it as 'changes in knowledge, understanding, skills and attitudes brought about by experience and reflection upon that experience' (Brown, *et al.*, 1997:124). In the last three decades is has been common place to interpret how learning occurs through a constructivist lens. Human beings have a powerful urge to make sense of what they hear or read (Donaldson, 1987). Vygotsky (1978) proposed that children constructed their own learning, but that this learning could be helped by adult intervention. He was aware that learners who were unable to perform tasks or solve problems alone often succeeded when an adult helped them. He used the term *zone of proximal development,* to describe 'the distance between the actual developmental level as determined by independent problem solving, and the level of potential development as determined through problem solving under adult guidance or in collaboration

Figure 1. Principles of learning (Hartley, 1998)

Instruction should be well-organized
Instruction should be clearly structured
The perceptual features of the task are important
Prior knowledge is important
Differences between individuals are important as they will affect learning
Cognitive feedback gives information to learners about their success or failure concerning the task at hand.

with a more capable peer' (Vygotsky, 1978:182). For Vygotsky, this co-operatively achieved greater success was crucial to both the learning and development of learner (Wood, 1988). Application of this constructivist model of learning to HE means that the intervention of the tutor may involve a kind of scaffolding process that enables the student to solve a problem or achieve a goal which would be beyond their unassisted efforts (Wood *et al.,* 1976).

The Collins Dictionary defines learning as "knowledge gained by study, instruction or scholarship; the act of gaining knowledge; any relatively permanent change of behaviour that occurs as a direct result of experience." Learning is defined by Birkenholz (1999) as a change in behaviour demonstrated through implementing knowledge, competences, skills, or practices derived from an educational experience. Hartley (1998) present six principles of learning, see figure xxx, that for them offer a strong position for the embedding of a constructivist approach which can be incorporated into an 'active' learning approach. Through the lens of an educator, learning involves working with learners along the learning process where the 'learning' includes all those things that are engaged with to make it happen. As an end result, learning occurs when new found information is understood and incorporated into an aspect of life. For example, working with an audience that have no knowledge or skill in public speaking, one outcome would be to see those learners gain knowledge in this area and to actually implement the new skills; this would exemplify true learning.

Student Centered Learning Environments

As referred to earlier, it has become fashionable and appropriate to talk about constructivism in relation to higher education learning, teaching and assessment.

Constructivist learning theory, a philosophy of learning that relates to the way in which individuals learn to make sense of the world by the development of learning constructs or mental representations of knowledge. These representations are created subjectively by using higher level reflective processes and are based on prior learning or knowledge. For effective learning to take place new information must be linked to frameworks of prior knowledge in order to make sense and produce meaning. The learner brings with them mental representations of knowledge based on their different levels of experience. The concept of constructivism has its roots in classical antiquity, going back to the work of Socrates. The Piagetian ideas of assimilation, accommodation and cognitive equilibrium, along with the work of Kelly (1955) on personal construct theory and Bruner's (1960) framework for instruction, have played very important roles in the development of the constructivist pedagogy. At the core of this theory is the relationship between the learner and the content, the view of learning is one in which the learners 'use their own experiences to construct understanding that makes sense to them, rather than having understanding delivered to them in already organised form' (Kauchak & Eggen, 1998:184). Grabinger (1996) proffers three notions that permeate this student centered approach; firstly, that knowledge acquisition is an active process in which the learner makes sense of the world as opposed to merely assimilating facts. Secondly, the learner internalizes new knowledge in personal ways through creating a relationship with existing knowledge and so enabling application. Thirdly, knowledge has a cultural aspect that relies on social negotiation and collaboration to give shared meanings.

Though the emphasis is on the learner, constructivism does not dismiss the active role of the lecturer/teacher or the value of the expert knowledge. However, their roles now change to that of facilitator. They will in essence lecture less and be much more as Weimer puts it 'around the classroom rather than in front of it (Weimer, 2002:14) but also employ more appropriate and authentic teaching strategies. Learning will become the important element and the learner will move from a passive recipient of information to active participant in the learning process which, as Fosnot (1996) suggests 'requires invention and self-organisation on the part of the learner'. Those who employ a constructivist philosophy in their teaching strategy recognise the four characteristics of constructivist teaching identified by Kuachak and Essen: learners construct their own understanding, new learning depends on current understanding, learning is facilitated by social interaction, and meaningful learning occurs within authentic learning tasks (Kauchak & Eggen, 1998:185).

If we desire to assist learners to take full responsibility for their own learning, develop their higher order thinking skills, and engage fully and actively in their learning, we need a grounded understanding of the student centered teaching practices we should implement. While some might be inspired by a constructivist philosophy, there is no doubt that there are many teachers/lecturers who, without

Figure 2. The continuum of learning and teaching in BE education

Traditional	Student Centred Learning
(didactic)	(participatory)

knowing or without having been informed of the theory, are providing the students in their care with opportunities for constructivist learning. The comments of von Glaserfeld ring through here:

Constructivism does not claim to have made earth-shaking inventions in the area of education; it merely claims to provide a solid conceptual basis for some of the things that, until now, inspired teachers had to do without theoretical foundation. (von Glaserfeld, 1995)

One of the current contradictions of the HE system globally is the drive towards becoming a 'mass' system (a system that is increasingly attracting a quantitatively and qualitatively diversified student population) while at the same time encouraging and implementing a learning and teaching environment that puts the learner at the centre of the learning experience. Taras (2002) points to the undermining of educational credibility among both staff and students in British universities as identified by the contradictions between the aims and pedagogic processes. She points to the protocols of both internal and external quality assurance deterring academics from developing and implementing innovations, while students are frustrated and disillusioned by what she terms the 'perennial pressures of summative grades' (Taras, 2002:501). SCL has in theory the potential to bring about greater learner involvement and participation and diminish this frustration. Much has been done, most notably in the last decade, to develop the theoretical grounding, practical support and assistance for student learning to flourish.

Kember (1997) articulates two broad orientations in teaching; the teacher-centered or content orientated conception and the student centered or learning orientated conception. Harden and Crosby (2000) refer to the teacher centered learning strategies as being focused on the teacher as the transmitter of knowledge from expert to novice as an example. In contrast, they describe student centered learning as being focused on the students' learning emphasising the concept of the learner 'doing' (Harden & Crosby, 2000:335). Student centered and teacher centered practices are often viewed as representing opposite ends of the teaching philosophy spectrum. See Figure 2.

Lea et al (2003:322) in their research of HE students' attitudes to student-centered learning summaries SCL as follows:

- The reliance on active rather than passive learning
- An emphasis on deep learning and understanding rather than surface learning
- Increased responsibility and accountability on the part of the student
- An increased sense of autonomy in the learner
- Interdependence between learner and lecturer
- Mutual respect within the learner lecturer relationship
- A more reflexive approach to the teaching and learning process on the part of the teacher and learner.

While the term student centered learning (SCL) is widely used in both educational literature about learning and teaching in HE and by those concerned with the implementation of the learning environment (lecturers and administrators alike), many phrases have been associated and linked to SCL and these include self-directed learning, experiential learning, participatory learning, humanistic and so on. In fact, the following terms have been associated with SCL by the many individual authors on the subject:

- Learner focused learning
- Active learning
- Autonomous learning
- Independent learning
- Collaborative learning
- Experiential learning (Dewey)
- Authentic learning
- Problem based learning
- Constructivist learning
- Enquiry (Socrates)

Although we consider the concept of SCL to be a relatively new one, particularly in Higher Education, interest in it has been long-standing among educators in all realms of education. The research, policy and practice has continued to grow and is based on a pedagogy that puts student learning at the centre and recognises and is supportive of individual learner needs, capacities/abilities, interests and experiences. However, it is appropriate to acknowledge that there is not universal agreement on the definition of SCL. Farrington has noted that there is 'considerable disagreement and confusion about what student centered learning actually is' (1991:16).

Brandes and Ginnis tell us that the claim for the term SCL goes to Carl Rogers (Rogers 1965) as it was he who invented it. It graphically describes what it is we are hoping to achieve, a system of providing learning which has the student at its heart (Brandes & Ginnis, '96:01). Gibbs (1992) offers a useful definition of student centered learning, he states that SCL 'gives students greater autonomy and control over choice of subject matter, learning methods and pace of study (p23). This view identifies three core characteristics of SCL and promotes the idea that each learner should have input into what is to learned, how it might be learned and when it is learned. Cowan (2005) has expressed it in terms of four characteristics, namely, pace of learning, method of learning, content of learning and the assessment of that learning.

A very important implication of the Gibbs and Cowan approaches to SCL, as has been alluded to already, is that there is a need for the learner to assume a high level of responsibility in the learning process and to be actively involved in managing their learning. This paradigm shift, which, while evolving slowly, has begun to take a greater effect in contemporary pedagogy, takes the responsibility from the teacher/lecturer and places it firmly in the hands of the learner. This requires learners to be, what de la Harpe et al (1999) have expressed as 'effective learners'.

Cannon and Newble (2000:16) provide us with a well defined definition for SCL and they define it as:

Ways of thinking and learning that emphasise student responsibility and activity in learning rather than what the teachers are doing. Essentially SCL has student responsibility and activity at its heart, in contrast to a strong emphasis on teacher control and coverage of academic content in much more conventional, didactic teaching.

Student Centered Learning shifts the power of learning to the learner, it offers the concept of the student's choice in their education and offers the opportunity for the learner to be a more active participant in their learning as opposed to a traditional passive recipient from the 'font of knowledge'. It affords the learner the opportunity to engage in deep learning which may be reflected upon as they progress. The Canon and Newble definition for this author covers the realm of SCL very appropriately; however, it should be would further add that implicit in the definition is that learners should be consulted on the learning, teaching and assessment process in the appropriate ways for where the learner is at on the continuum. Examples of this practice exist and continue to be expanded in the Built Environment, where teachers/lecturers, for example, are engaging in Problem Based Learning (PBL), use of portfolio assessment and in many cases have negotiated with the learners how this might be best advanced.

Changing the Experience of the Learner

Perhaps one of the important drivers for attempting to implement active learning and more student centered approaches in higher education is the changing nature of the student population. There will always be some students who will thrive and prosper without attending many lecturers. However, with widening participation, it is no longer the elite who are in HE nowadays, and our class groupings now include many students who actually need the focus which a good shared experience can give them, helping them to see the bigger picture into which they need to fit the smaller bits. Back as far as 1996, Cross proffered three conditions for learner success in student learning and these are:

1. High expectations,
2. Student participation and involvement, and
3. Assessment and feedback.

Students should have high expectations in what they want to achieve and they should be encouraged and supported in their participation and involvement through what Sadler (1989) refers to as the setting of 'hard goals'. Responsible learners are defined by Scharle and Szabó as those 'who accept the idea that their own efforts are crucial to progress in learning' (2000:03); they are aware of their responsibilities while understanding the consequences on their progress if they fail to involve themselves. Autonomous learning, while like responsibility, requires active involvement, it is regarded as the freedom and ability to manage learning. Autonomy and responsibility are attitudes that learners possess to varying degrees along with personality traits, preferred learning styles and cultural attitudes influencing the development of both autonomy and responsibility.

The responsibility for learning in the context of a SCL approach has implications for and involves action on the part of the learners. They must accept responsibility for their learning and this involves developing the intellectual maturity, the required learning skills and in particular the necessary awareness to function as independent and autonomous learners. As referred to earlier, in a truly SCL approach, it is the responsibility of the 'learned' to take explicit actions that will motivate and enhance student learning. While some pioneering academics have engaged in the full implementation of a truly student centered approach others have been slow to adopt this approach and have cited the lack of maturity of the learners they meet as the main barrier. Also argued is the fact that the experience of those entering HE from the second-level education system, mainly a rote learning system, tends to militate against this in that students tend to be overly reliant on the support of academics. Initial research by Fazey (1996) seems to suggest otherwise. It indicates that first year university students have a higher perception of autonomy than either their second of third year counterparts and that it seems to decrease as they progress through college. Taras (2002) suggests an explanation for this could perhaps be that students are being provided with mixed messages and that there seems to be a disconnect between the rhetoric and reality of what is said and what is done.

The BE academic community must challenge students to take responsibility and ownership of their learning and while acknowledging that instances of the traditional teacher led model still exist, there are ever increasing examples of Schools who have deviated from that traditional model. These have developed and implemented 'key skill modules' where learners now engage in constructivist experiences with exposure to units that engage them as autonomous independent learners. Thus the abilities in each learner to take charge of their own learning are built up. From informal discussions with those involved indications are that a positive experience for both the learner and themselves has been experienced. Learners further on in the programme have given positive feedback that indicates their self-confidence and decision making skills have increased greatly and this has impacted significantly on later learning.

Changing the Experience of the Learned

Thirty-five or so years ago, HE functioned via the myth that university and college academics only needed expertise in their disciplines in order to be good teachers. However, in the 1970s this myth begun to be challenged from sectors both within and outside higher education. Since that time higher education has expanded and diversified with formidable demands and expectations being placed on colleges and universities, both nationally and internationally. Changes included increased pressures through increased compliance required by government on quality assurance, costs, results and retention.

Various research studies (for example, Powell & Anderson, 2002; Anderson & Helms, 2001; Stronge & Tucker, 2000; Mendro, 1998) have emphasised the importance of the role of the teacher/lecturer in the successful implementation of an educational programme. It is also recognised, for academic staff there are increased internal pressures through growing quality assurance procedures, additional administrative requirements and the pressure to develop and in some instances strengthen their research profile. Because learning in educational institutions has been traditionally dominated and controlled by adults (i.e. lecturers, teachers, etc), students tend to allow others make the decisions about their learning. Although our teaching philosophy may purport to graduating students who are responsible citizens, capable of participating thoughtfully in a democracy, our educational practices often tend to foster dependence, passivity and a 'tell me what to do and think' attitude in them despite the good intentions of academics. In particular, it has obstructed the attainment of one of the common goals of higher education – that of students becoming autonomous learners who take responsibility for their own learning. To what extent do we actually engage our students in meaningful student centered learning, and to what extent should we? Interestingly, some teacher/lecturers have resisted the change preferring to continue with the traditional teacher led approaches. Weimer (2002) suggests this is because the traditional approach requires less of them and SCL approaches to learning, teaching and assessment take a considerable amount of effort on the part of the teacher/lecturer. While teaching large classes is often the norm in HE and the flexibility to use other methodologies other than lectures is not readily accessible, the lecture should be arranged so that it causes learning to happen. We need to inspire students with our enthusiasm for the discipline and use the large group as an opportunity to help students see what exactly we expect of them, so they know what they are supposed to be trying to achieve.

In approaching the design of curricula it is important to be aware of the theories that underpin learning, taking cognizance of the fact that there is no universal way of learning. Biggs (1999) proposes the 'constructive alignment' model as an approach to curriculum design that maximises the opportunity for quality learning. The traditional model of curriculum, teaching and assessment, where the curriculum entailed a list of topics, the teaching involved lectures and the assessment involved a summative examination, suffers from the basic problem that, with this model there is no intrinsic link between the curriculum, the teaching or the assessment. To ensure and encourage deep engagement from the learner the Biggs constructively aligned model is appropriate, where the process should be aligned from the forming of appropriate level learning outcomes with the most suitable choice of learning/teaching methods matched with authentic assessment of the intended learning outcomes. This approach has been at the core of the realignment of both programme and module curricula in many of the BE programmes across the world. What is often neglected is taking time to engaged newly recruited colleagues as to the beginning and emergence of such realignment.

Even if one is convinced about the validity of SCL and has the understanding of the underlying theories and principles, the question of how to go about actually changing things in the learning environment in which you operate, may still remain. Ownership is at the heart of SCL and developing and sharing best practice is key here. The provision of support structures so that actual 'pedagogical transformation' (Major & Palmer, 2006) can take place is vital as it is important to embed appropriate methodologies that allow students engage in meaningful learning. Major and Palmer identify developments that have helped in the advancement of academic development in their institution including: release time for module development, the inclusion of guest speakers (to include subject matters experts from well renowned institutions) academic development workshops, monetary support for appropriate resources, support for travel to engage with and experience the use of current best practice. Some practical work done by the learning and teaching development teams particularly in meeting the requirements of the Bologna Process has included working with programme teams, in preparation for modularisation, on the application of 'constructively aligned' (Biggs, 1999) module development, support and direction in the application of more formative assessment methodologies. Many examples come to mind and these include professional development for academics in different areas.

Shuell (1986: 429) stated that:

If students are to learn desired outcomes in a reasonably effective manner, then the teacher's fundamental task is to get students to engage in learning activities that are likely to result in their achieving those outcomes.

Biggs (2003) advanced Shuell's ideas to include assessment in an aligned system of teaching. This alignment of outcomes, teaching methods and assessment is central to facilitating successful learning for students. He goes on to suggest that teaching should be based on the constructivist theory (CT) of learning where ever possible. The CT theory premise is that meaning is not imposed or transmitted by direct instruction, but that meaning is created by the learners' through learning activities. Constructivism also, suggests that we learn by fitting new understanding and knowledge into, with, extending and supplanting, old understanding and knowledge. What students will construct from a learning encounter depends on 'student factors' such as prior knowledge, ability and motivation, the 'teaching context' such as objectives, assessment, ethos, teaching, institutional procedures, and the teaching/learning activities. Additions to knowledge, in the sense of accumulated 'facts', may sometimes be possible without any substantial transformation, but higher order learning involving understanding, can usually only happen when there is conceptual change to incorporate new understanding (Fry *et al.*, 2003). Thus Biggs (1999) embraced the notion of 'constructive alignment' in programme and module design, which he conceived as a "marriage between a constructivist understanding of the nature of learning, and an aligned design for teaching" (Biggs, 1999:26).

In a constructively aligned approach there are:

- Clear outcomes which state the level of understanding required (more than a list of topics)
- Learning and teaching activities that can create the outcomes to be met
- Assessment tasks that address the outcomes.

In a constructively aligned programme or module (unit of learning), the students themselves do the real work while the academic simply creates the environment so that it will happen.

While the 'constructively aligned' approach is being advocated by many, there are some who oppose it and for example it appears s to present a method by which Hussey and Smith (2002) can draw up the learning outcomes to which they aspire, being 'more realistic and conducive to educational purposes'. But rather than acknowledge this, the authors instead seize on the term 'entrapped', declaring they are 'wary of creating conditions that entrap the students' and that 'establishing [a context] which requires their passive adherence appears antithetical to the processes of learning and education' (p. 361). Hussey and Smith (ibid) also refer to learning outcomes as 'observable products of the learning process'. In making such a statements it seems they entirely misrepresent Biggs' philosophy, which sees 'the learner as central in the creation of meaning, not the teacher, as the transmitter of knowledge' (Biggs, 1996, p. 348).

Ecclestone, in her paper examining the implications of outcome-based assessment in HE, makes a key distinction between two approaches:

- Outcomes expressed as broad aims for teaching and learning… a useful aid in curriculum planning… and for making assessment criteria more rigorous and accessible to learners.
- Outcome-based assessment… expressed as detailed, standardised definitions of learning outcomes, accompanied by similarly specific assessment criteria. (ecclestone, 1999, pp. 30-31).
- Hussey and smith critique consistently focus on the latter, where they seem unshakeable in their conviction that by choosing to accommodate the interests of the students, to fully exploit the opportunities for deeper learning, an educator by design reduces the chances of achieving the pre-specified outcomes.

Because CL is relatively new to many institutions, teachers must work with their supervisors to make sure they are aware of the techniques as well as reasons for using them. Supervisors who are used to seeing orderly, quiet classes with students listening to a lecture need to be informed about the nature of CL classes, where students talk in groups and socialize while they work on task. A common concern among CL teachers is that their supervisors will not understand what is taking place in the class and that poor evaluations will result. This problem can be avoided by the institution's providing training for administrators, and by the teachers working closely with their department heads and other supervisory personnel.

To the early career academic what would be useful is to experiment with a more active and collaborative learning contexts and what could be useful in planning for that activity is to build in a framework of active learning structure to the learning environment. Figure 3 maps out a potential active learning framework that could be considered and where elements of the framework would be adopted to create an environment that would be conducive to active learning. (See Figure 3 below). It would not be necessary to include all elements of the framework at one time but what would be useful is to explore elements that have to date not be addressed.

Active learning processes include class warmup activities, name recognition games and group building activities, and group processing. Students work in pairs

Figure 3. Adaptable active learning framework

1. DEVELOP HIGHER LEVEL THINKING SKILLS
2. PROMOTE STUDENT-ACADEMIC INTERACTION AND FAMILIARITY
3. FOSTERS TEAM BUILDING AND A TEAM APPROACH TO PROBLEM SOLVING WHILE MAINTAINING INDIVIDUAL ACCOUNTABILITY
4. BUILD IN SELF ESTEEM DEVELOPMENT APPROACHES
5. ENHANCE STUDENT SATISFACTION WITH THE LEARNING EXPERIENCE
6. PROMOTE A POSITIVE ATTITUDE TOWARD THE SUBJECT MATTER
7. DEVELOP ORAL COMMUNICATION SKILLS OF STUDENTS
8. DEVELOP SOCIAL INTERACTION SKILLS OF STUDENTS
9. STUDENTS ARE TAUGHT HOW TO CRITICIZE IDEAS, NOT PEOPLE
10. CREATE AN ENVIRONMENT OF ACTIVE, INVOLVED, EXPLORATORY LEARNING
11. STUDENT DEVELOP RESPONSIBILITY FOR AND TO EACH OTHER
12. ENCOURAGE DIVERSITY OF UNDERSTANDING AND THE SHARING OF THAT UNDERSTANDING
13. ENCOURAGE STUDENT RESPONSIBILITY FOR LEARNING
14. INVOLVE STUDENTS IN DEVELOPING CURRICULUM AND CLASS PROCEDURES
15. STUDENT EXPLORE ALTERNATE PROBLEM SOLUTIONS IN A SAFE AND SUPPORTIVE ENVIRONMENT
16. STIMULATE CRITICAL THINKING AND HELPS STUDENTS CLARIFY IDEAS THROUGH DISCUSSION AND DEBATE
17. ENHANCE SELF MANAGEMENT SKILLS
18. FITS IN WELL WITH THE CONSTRUCTIVIST APPROACH
19. ESTABLISH AN ATMOSPHERE OF COOPERATION
20. BUILD MORE POSITIVE HETEROGENEOUS RELATIONSHIPS
21. ENCOURAGE ALTERNATE STUDENT ASSESSMENT TECHNIQUES
22. FOSTER AND DEVELOP INTERPERSONAL COMMUNICATION
23. MODELLING OF PROBLEM SOLVING TECHNIQUES BY STUDENTS' PEERS
24. SET HIGH EXPECTATIONS FOR STUDENTS AND TEACHERS
25. PROMOTE HIGHER ACHIEVEMENT AND CLASS ATTENDANCE
26. STUDENTS FOCUS AND STAY ON TASK MORE AND ARE LESS DISRUPTIVE
27. GREATER ABILITY OF STUDENTS TO VIEW SITUATIONS FROM OTHERS' PERSPECTIVES (DEVELOPMENT OF EMPATHY/ UNDERSTANDING)
28. CREATE A STRONGER SOCIAL SUPPORT SYSTEM
29. CREATE A MORE POSITIVE ATTITUDE TOWARD
30. PROMOTE INNOVATIVE TEACHING AND CLASSROOM TECHNIQUES
31. CLASSROOM SHOULD RESEMBLE REAL LIFE SOCIAL AND EMPLOYMENT SITUATIONS
32. STUDENTS PRACTICE MODELLING SOCIETAL AND WORK RELATED ROLES

or larger groups depending upon the task at hand. Group work on content takes many forms, including pairs or groups working on individual questions, problem assignments, projects, study activities, group tests etc. Classes are interesting and enjoyable because of the variety of activities available for use by the teacher. In fact, collaborative learning effectively addresses the "Sesame Street" syndrome in which modern students are used to being exposed to information in short, entertaining sessions. These same students are also used to high tech computer systems which deliver material in a variety of ways including video, text, graphical illustrations, and interactive systems. The key is that Collaborative learning effectively matches or exceeds the above approaches to learning by actively involving every student.

In a more collaborative learning environment students may be assigned roles in order to build interdependence within the groups. Roles such as reader, recorder, reporter, materials handler, time keeper, sceptic/challenger and others may be rotated among group members for each new assignment or project (Johnson and Holubec 1984). The learner is thus encouraged to develop and practice their skills, ones which will be needed to function in society and the work world (Houston 1992) but also with the given context to consolidate that learning. These competences and skills would include leadership, information recording, communication of results orally and in writing, challenging ideas in a constructive manner, obtaining and distributing materials and information to group members, encouraging member participation, brainstorming, meeting deadlines, etc. (Sandberg 1995). Wlodowski (1985) observes that, "If students realize the direct applicability of classroom small group problem-solving to their own lives, motivation to learn will show a marked increase." Building strong social characteristics within the learner can be practiced in a risk free environment with support and experience from the teacher but also form peers. This offers an environment where meaningful learning should be achieved.

FUTURE RESEARCH DIRECTIONS

In 1946 the social psychologist Kurt Lewin wrote a paper advocating a 'type of action- research [sic]' that would lead to social action. 'Research that produces nothing but books,' he argued, 'will not suffice' (Lewin 1946: 35). Instead Lewin proposed a cyclical, iterative approach to research involving planning what was to be done, taking action and fact-finding about the results. Lewin's ideas have since become one of the key influences in what is now known as action research. Over time, action research has taken different directions and for the embedding of good practice, that must be reflected upon a future direction proposed here is that Built Environment academics become action research and build on their practice with continuous improvement through the refinement of their practice. This can then be share with the community; a community of practice as it were!

The close relationship between knowledge acquisition and action; action research is 'research in action rather than research about action' (Coghlan and Brannick 2010: 4). Action is taken to improve practice and the research generates new knowledge about how and why the improvements came about.

Action research can be conducted as a collaborative partnership between the researcher (the lecturer) and a group in an organisation or community who participate in the process of the action research. Research proceeds as a cycle of joint planning, action, observation and reflection, where the reflection phase paves the way for further cycles of planning, acting, observing and reflecting in a spiral of learning (Figure 1). Different writers have used different terminology for the steps in the action research cycle. Coghlan and Brannick (2010: 8), for example, label them 'constructing', 'planning action', 'taking action' and 'evaluating action'. The results are shared amongst participants and for action research in an academic context the output typically also includes a public report such as a dissertation or thesis.

The output is 'actionable knowledge' (Coghlan 2007: 293) that is useful to both the practitioner and academic communities. There is certainly scope for this within the BE community.

CONCLUSION

For the early career academic trying to adapt/ adopt learning, teaching and assessment strategies while trying to address the myriad of aspects that are required during those early years requires a commitment to be bold, experiment and reflect on those early efforts. Having the learner at the centre of the experience is the key to success. Active learning practitioners would say that all active/collaborative learning is about building learning communities. However, we use of the term learning community is often referred to in a broader but more specific sense, in terms of intentional reconfiguration of the programme curriculum. In more recent times, there are a scattering of built environment programmes who have researched and recognized that deep-seated structural factors weaken the quality of undergraduate learning and inhibit the development of community. The response by these schools has been to develop learning communities, a "purposeful restructuring of the curriculum to link together courses so that students find greater coherence in what they are learning and increased interaction with faculty and fellow students" (Gabelnick et al, 1990). As such, these learning communities have a structured system and within that facilitate the practice of active learning. Learning community curriculum structures vary from campus to campus. Their intent serves many different purposes, but have two common intentions. They attempt to provide intellectual coherence for students by linking classes together and building relationships between subject matter, or by providing the opportunity for active learning in the context of a discipline. Second, they aim to build both academic and social community for students by enrolling them together in a large block of course work underpinned by active learning tasks. Learning communities directly confront multiple problems plaguing under-graduate education: the fragmentation of general education classes, isolation of students (especially on large campuses or commuter schools), lack of meaningful connection- building between classes; the need for greater intellectual interaction between students and faculty; and lack of sustained opportunities for faculty development.

By altering the curricular structure to provide larger units of study, learning communities frequently provide more time and space for collaborative learning and other more complicated educational approaches. Small group workshops and book seminars are staples of most learning communities. Peer writing groups and team projects associated with labs and field work are also fairly common. Study groups emerge in learning communities, both intentionally and spontaneously. These programs provide a unique social and intellectual glue for students that results in high rates of student retention, increased student achievement and more complex intellectual development (MacGregor, 1991).

Creating an active learning environment can be a wonderfully rewarding opportunity but it is also full of challenges and dilemmas. Very few experienced active learning opportunities in our own undergraduate settings, and much of our graduate school training reinforced a teacher-centered, lecture-driven model of college teaching. So for this researcher/ educator, stepping out of the center and engaging students in group activity is hard work, especially at first. You are encouraged to start this in a structured way where the appropriate 'scaffolding' is provided for the learner.

Designing active and group learning activities requires a demanding yet important rethinking of a course syllabus, in terms of course content and time allocation. If some (or a great deal) of the classroom time is considered an important social space for developing understandings about course material, or if some of the out-of-class time is devoted to study groups or group projects, how should the rest of the class time (lectures, assignments, examinations) be best planned for? How does one ensure students are learning and mastering key competences, knowledge, skills and ideas in the course, while at the same time addressing all the material of the course? Teaching in where an active learning approach is preferred puts front and center the tension between the process of student learning and content coverage.

As we become more involved in using active/collaborative learning, we discover what radical questions it raises. Active learning goes to the roots of long-held assumptions about teaching and learning. Classroom roles change: both the learned and the learner take on more complex roles and responsibilities (Finkel and Monk, 1983; MacGregor, 1990). The classroom is no longer solo learned and individual learner - it becomes more an interdependent community with all the joys and tensions and difficulties that attend all communities. This degree of involvement often questions and reshapes assumed power relationships between the stakeholders, (and even between students and students), a process that at first can be confusing and disorienting (Romer and Whipple, 1990).

Not only is course content reshaped, so too is the definitions of student competence. Due to the public nature of active learning it makes demonstration of student learning very continuous and collaborative however, it does both complicate but enrich the assessment/ evaluation process.

What can be a challenge to the active learning environment at the classroom level are compounded by the traditional structures and culture of the academy, which continue to perpetuate a teacher-centered, transmission of information model of learning, teaching and assessment. The political economy of the academy is set up to front load the curriculum with large lower division classes in rooms immutably arranged for lectures, usually in classes limited to fifty-minute. Student-student interaction; extended, careful examination of ideas; the hearing-out of multiple perspectives; the development of an intellectual community - all these are hard to accomplish under these constraints.

What has really has propelled this educator into collaborative and active learning approach is the desire to motivate students by getting them more actively engaged. Nonetheless, wanting to be a facilitator of active/collaborative learning and being good at it are very different things. As with all kinds of teaching, designing and guiding group work takes time to learn and practice. And for students, learning to learn well in groups provides challenges and does not happen overnight. The recommendation to those early career academics is to start with modest efforts, work with those colleagues interested in providing a positive experience to learners. Learn from each other through sharing your experiences and best practice.

At their best, active learning environments stimulate both students and teachers. In the most authentic of ways, the active learning process models what it means to question, learn and understand in concert with others. Learning actively demands responsibility, persistence and sensitivity, however the result can be a community of learners in which everyone is welcome to join, participate and grow. Where learning is the goal!!!

REFERENCES

Biggs, J. (1999). Enhancing teaching through constructive alignment. *Higher Education, 32*(3), 347–364. doi:10.1007/BF00138871

Biggs, J. (2007). *Teaching for Quality Learning in University*. Open University Press.

Birkenholz, R. (1999). *Effective adult learning*. Danville, IL: Interstate Publishers, Inc.

Brandes, D., & Ginnis, P. (1996). *A Guide to Student Centred Learning*. Simon & Schuster.

Brown, G., Bull, J., & Pendlebury, M. (1997). *Assessing student learning in higher education*. London: Routledge.

Bruner, J. (1970). *The Process of Education*. Cambridge, MA: Harvard University Press.

Cannon, R., & Newble, D. (2000). *A Guide to Improving Teaching Methods: a handbook for teachers in universities and colleges*. Kogan Page.

Coghlan, D. (2007). Insider action research: Opportunities and challenges. *Management Research News, 30*(5), 335–343. doi:10.1108/01409170710746337

Coghlan, D., & Brannick, T. (2014). *Doing Action Research in Your Own Organization*. Sage Publications London.

Cowan, J. (2005). *Student-centered learning*. Paper presented at DIT Learning and Teaching Showcase, Dublin Institute of Technology.

Cross, K. P. (1996). Improving teaching and learning through classroom assessment and classroom research. In Improving student learning using research to improve student learning. Academic Press.

de la Harpe, B., Kulski, M., & Radloff, A. (1999). How best to document the quality of our teaching and our students' learning? In K. Martin, N. Stanley, & N. Davison (Eds.), Teaching in the Disciplines/ Learning in Context (pp. 108–113). Academic Press.

Donaldson, M. (1987). The origins of inference. In J. Bruner & H. Haste (Eds.), *Making Sense. The Child's Construction of the World*. London: Routledge.

Ecclestone, K. (1999). Empowering or Ensnaring? The Implications of Outcome-based Assessment in Higher Education. *Higher Education Quarterly, 53*(1), 29–48. doi:10.1111/1468-2273.00111

Farrington, J. (1991). Student-centred learning: Rhetoric and reality*? Journal of Further and Higher Education*, *15*(3), 16–21. doi:10.1080/0309877910150302

Fazey, D. (1996). Individual differences in the autonomy related psychological characteristics of undergraduates. In G. Gibbs (Ed.), *Improving student learning using research to improve student learning* (pp. 187–199). Oxford, UK: Oxford Centre for Staff Development.

Finkel, D., & Monk, G. (1983). Teachers and Learning Groups: Dissolution of the Atlas Complex. In C. Bouton & R. Y. Garth (Eds.), *Learning in Groups. New Directions for Teaching and Learning*. Jossey-Bass. doi:10.1002/tl.37219831411

Fosnot, C. T. (1996). *Constructivism: Theory, Perspectives and Practice*. New York: College Teacher Press.

Fry, H., Ketteridge, S., & Marshall, S. (1999) Understanding student learning. In A Handbook for Teaching and Learning in higher Education. Enhancing Academic Practice. London: Kogan Page.

Gabelnick, F., MacGregor, J., Matthews, R., & Smith, B. (1990). *Learning Communities: Creating Connections Among Students, Faculty and Disciplines*. San Francisco: Jossey Bass.

Gibbs, G. (1992). *Improving the quality of student learning*. Bristol, UK: TES.

Grabinger, R. S. (1996). Rich environments for active learning. In D. H. Jonassen (Ed.), *Handbook for research for educational communications and technology*. New York: MacMillan.

Harden, R. M., & Crosby, J. (2000). AMEE Guide No 20: The good teacher is more than a lecturer-the twelve roles of the teacher. *Medical Teacher*, *22*(4), 334–347. doi:10.1080/014215900409429

Hartley, J. (1998). *Learning and Studying. A research perspective*. London: Routledge. doi:10.4324/9780203270189

Hussey, T., & Smith, P. (2002). The trouble with learning outcomes. *Active Learning in Higher Education*, *3*(3), 220–233. doi:10.1177/1469787402003003003

Johnson, D., & Johnson, R. (1984) Structuring Groups for Cooperative Learning. *Journal of Management Education*, *9*(4).

Kauchak, D. P., & Eggen, P. D. (1998). *Learning and Teaching: Research Based Methods*. Boston: Allyn and Bacon.

Kember, D. (1997). A re-conceptualisation of the research into university academics conceptions of teaching. *Learning and Instruction*, *7*(3), 255–275. doi:10.1016/S0959-4752(96)00028-X

Lea, S. J., Stephenson, D., & Troy, J. (2003). Higher education students' attitudes to student-centred learning: Beyond the 'educational bulimia'? *Studies in Higher Education*, *28*(3), 321–334. doi:10.1080/03075070309293

Lewin, K. (1935). *A Dynamic Theory of Personality*. New York: McGraw Hill.

Mac Gregor, J. (1991). *What Difference Do Learning Communities Make?* Olympia, WA: Washington Center for Undergraduate Education, The Evergreen State College.

MacGregor, J. (1990). Collaborative Learning: Shared Inquiry as a Process of Reform. In M. Svinicki (Ed.), *The Changing Face of College Teaching. New Directions for Teaching and Learning, No. 42*. San Francisco: Jossey- Bass. doi:10.1002/tl.37219904204

Major, C. M., & Palmer, B. (2006). Reshaping teaching and learning: The transformation of faculty pedagogical content knowledge. *Higher Education*, *51*(4), 619–647. doi:10.100710734-004-1391-2

Mendro, R. L. (1998). Student achievement and school and teacher accountability. *Journal of Personnel Evaluation in Education*, *12*, 257–267.

Peel, D. (2010) Editorial: Making an Impact in the Built Environment? *Journal for Education in the Built Environment*, *5*(1), 1-6.

Rogers, C. (1965). *Client-centred Therapy*. Houghton-Mifflin.

Romer, K., & Whipple, W. (1990). Collaboration across the Power Line. *College Teaching*, *39*(2), 66–70. doi:10.1080/87567555.1991.9925490

Sadler, D. R. (1989). Formative assessment and the design of instructional systems. *Instructional Science*, *18*(2), 119–144. doi:10.1007/BF00117714

Sandberg, K. E. (1995). Affective and cognitive features of collaborative learning. Review of Research and Developmental Education, 6(4).

Scharle, A., & Szabó, A. (2000). *Learner Autonomy: A guide to developing learner responsibility*. Cambridge University Press.

Scott, L., & Fortune, C. (2009). Promoting student centered learning: the use of portfolios in the assessment of learning on an undergraduate construction management program. *Proceedings of the Associated Schools of Construction Conference*.

Scott, L., & Fortune, C. (2010). An Investigation of Assessment Practices in Built Environment Education – the Views of Senior Academics. *Proceedings Associated Schools of Construction 46th Annual International Conference.*

Scott, L., & Fortune, C. (2011). Reshaping assessment in Built Environment undergraduate programs - the challenge. *Proceedings Associated Schools of Construction 47th Annual International Conference.*

Scott, L., & Fortune, C. (2012). Towards Assessment Practices 'for' Learning in Irish Built Environment Higher Education. *Journal for Education in the Built Environment.*

Scott, L., Harvey, J., & Hayes, N. (2007). *Enhancing Student Learning through Assessment and Feedback A Toolkit Approach.* Dublin Institute of Technology.

Shuell, T. (1986). Cognitive conceptions of learning. *Review of Research in Education, 19*, 405–450.

Stronge, J. H., & Tucker, P. D. (2000). *Teacher evaluation and student achievement.* Washington, DC: National Education Association.

Taras, M. (2002). Using assessment for learning and learning from assessment. *Assessment & Evaluation in Higher Education, 27*(6), 501–510. doi:10.1080/0260293022000020273

Von Glaserfeld, E. (1995). A constructivist approach to teaching. In L. Steffe & J. Gale (Eds.), *Constructivism in education* (pp. 3–16). Lawrence Erlbaum Associates.

Vygotsky, L. S. (1978). *Mind in society: the development of higher psychological processes.* Harvard University Press.

Wlodkowski, R. J. (1985). *Enhancing Motivation to Learn.* San Francisco: Josey-Bassey.

Wood, D. (1988). *How children think and dream.* Oxford, UK: Basis Blackwell.

Wood, H. A., & Middleton, D. J. (1976). An experimental evaluation of four face-to-face teaching strategies. *International Journal of Behavioral Development, 1*(2), 131–147. doi:10.1177/016502547800100203

Wright, L., & Jones, A. (2007). *Good Practice and Innovation in the Assessment of Students in Built Environment Subject in Higher Education.* Liverpool John Moores University.

KEY TERMS AND DEFINITIONS

Active Learning: Active learning refers to a form of learning in which the teaching approach strives to involve students in the learning process more directly than in other methods. In an active learning environment, the learners participate in the process and learn by doing.

Assessment: Assessment can be defined as the systematic and ongoing method of gathering, analyzing, and using information from measured outcomes to improve student learning in terms of knowledge acquired, understanding developed, and skills and competencies gained.

Built Environment: The term refers to the human-made surroundings that provide the setting for human activity, ranging in scale from buildings to parks. In terms of built environment education it refers to the subject that define the built environment and include architecture, surveying, construction management and technology, urban planning, and so on.

Collaborative Learning: Collaborative learning is a situation in which two or more people learn or attempt to learn something together. Unlike individual learning, people engaged in collaborative learning capitalize on one another's resources and skills.

Diagnostic Assessment: Assessment designed to identify skills, knowledge and other attributes which can be used to decide on specific pathways of study, or difficulties in learning which require support.

Evaluation: The process of observing and measuring a thing for the purpose of judging it and of determining its "value," either by comparison to similar things, or to a standard.

Feedback: Information provided to learners about their work and progress.

Formative Assessment: Improvement-oriented assessment which is not marked for summative purposes.

Learning by Doing: Refers to a theory of education developed by American philosopher John Dewey. Dewey theorized that learning should be relevant and practical, not just passive and theoretical.

Learning Outcomes: Observable behaviors or actions on the part of students that demonstrate that the intended learning objective has occurred. Used to express intended results in precise terms.

Summative Assessment: Accountability-oriented assessment. The use of data assembled at the end of a particular sequence of activities, to provide a macro view of teaching, learning, and institutional effectiveness.

Chapter 2
A Construction Management Education Focus and Process Direction:
The Power of Focusing on Four Outcomes Using Formative Teaching, Learning, and Assessment

Matthew Steele Stevens
Western Sydney University, Australia

Jennifer E. Day
University of Melbourne, Australia

ABSTRACT

This chapter outlines culminates experience and thinking from two environments: 1) the construction contracting industry and 2) the built environment classroom – undergraduate and graduate. The authors' combined teaching experience covers more than two decades, but they do not pretend their knowledge is exhaustive. Others know more. Their desire is to add to their thoughts. They first address what the contractor desires of construction graduates. From there, they proceed to list focuses they believe important in sequential order for the teaching and learning process: 1) vocabulary, 2) conceptual frameworks, 3) iterative problem solving, 4) writing. After that, they describe their preferred delivery method: formative teaching, learning, and assessment.

DOI: 10.4018/978-1-5225-8452-0.ch002

INTRODUCTION

Construction education is a critical component to improving the construction industry. The newly graduated have been immersed in today's effective practices of leadership, management, and technology. Each one of us has out-of-date ideas, but new graduates bring much up-to-date thinking to the industry. They have "fresh eyes" and want to learn to be efficient in our challenging environment. Overall, they have a positive effect on the industry.

We believe that improving industry readiness of the newly graduated is an important discussion. In this Chapter, structured in a white paper style, we suggest a limited number of teaching, learning and assessment focus as suggested by our industry interactions and classroom experience. We discuss our reasoning and conclusions for recommending four general learning focuses along with an already credible learning, teaching, and assessment process commonly known as the Formative Approach (Bloom 1968). Additionally, we attempt to address common criticisms.

This Chapter outlines our culminative experience and thinking from two environments – 1) The Construction Contracting Industry and 2) The Built Environment Classroom – undergraduate and graduate. Our combined teaching experience covers more than two decades, but we do not pretend our knowledge is exhaustive. Others know more. Our desire is to add to their thoughts.

Our work with clients has given to us a useful understanding of the contractor's desires of newly hired graduates' skills. Between us, we have been engaged by more than 150 contractors. As part of our engagements, we have interviewed dozens of graduates whose titles include Assistant PM, Field Engineer, Project Coordinator, Safety Manager, and Project Accountant. Regardless of their final educational degree – there appears to be an opportunity to improve this transformative process.

We see construction education as a journey that is never static nor complete. Our intent is to prompt more thinking and discussion about this important industry subject. We will use undergraduate education for applying our assertions.

We list suggestions that can be implemented in any course - standard or elective. Our focus is not about content. The proposed approach concerns the process and focuses for bringing along the knowledge and skills of the aspiring construction contracting professional.

Contractors will hire the majority of any construction program's graduates. This discussion is part of the healthy evolution of how to increase the quality of those graduates. We are not calling for new requirements nor expanded curriculum, only a consolidation of instructional direction to what construction contractors appear to want.

We first address what the contractor desires of construction graduates, in "Construction Contractor Wants." From there, we proceed to list focuses we believe important, in sequential order for the teaching and learning process: 1) Vocabulary 2) Conceptual Frameworks 3) Iterative Problem Solving 4) Writing. After that, we describe our preferred delivery method: Formative Teaching, Learning, and Assessment.

The first focus listed is "Vocabulary" starts the process. Students are immersed in construction labels and definitions. As many know, this helps instructor and students (and eventual employers) understand each other's questions and answers. Precise language is a value among most construction executives. The last focus listed "Writing" finishes the process. A person with an emerging writing skill will grow his or her thinking and speaking performance. Looking at past undergraduate education, it is no secret that writing is a core part of graduate school and for good reasons.

The four focuses and our recommended delivery method accumulate. From the first focus onward, the emphasis on vocabulary continues. Accordingly, the second focus on conceptual frameworks continues onto the last course. We list the four and discuss the approach that binds these together – the Formative Approach.

Construction Contractors' Wants

We have discussed desired employee attributes and performance with our clients over the years. Many times, this is during our advising work.

- **Independence:** Almost all contractors we know want their people to work independently and be "low-maintenance". Eventually, after some specific company training, a graduate will be expected to find his or her answers without involving others unnecessarily. Also that the person can engage the client and others without endangering their company's business-to-business relationship.
- **Steadiness:** Many contractors we have worked with seek those who have a craftsperson or equipment operator mentality. That is a steady work ethic with a constant awareness of what they are seeing. They exhibit less procrastination and then, rushing at the end to complete tasks. Rushing leads to mistakes and expensive rework.
- **Thoroughness:** Most contractors want employees who pay attention to detail and do the job right the first time. Slow, thoughtful and right is better than fast, thoughtless, and wrong. Rework is the costliest of all construction activities. This slow ethic may annoy some in the industry; however, those who practice this start earlier with planning the project or assessing the problem. Furthermore, the best employees have a critical eye for detail and a questioning attitude of anything that is out of the normal. Again, the mentality of the able craftsperson or equipment operator is the approach.

- **Integrity:** The best construction employees admit when they have made a mistake, seek guidance as they need it, argue for fairness, and take responsibility for the safety of everyone. They stay away from "gray" areas, if possible. Educators can help by insisting on minimum standards for students to uphold. As we have previously written, it is the responsibility of the company to create a company culture where an employee can exercise integrity, but it is also important to select employees with the demonstrated capacity to act with integrity. "Gritty" thinking appears to be a personal attribute common among these kinds of employees.

Lastly, it is important to note that each contractor we know states that each graduate they hire will have to be trained regardless of where they earned their degree(s). All construction firms have unique approaches and processes to constructing work. Add to that, their culture, philosophy, and clients' mentality is not known to the newcomer.

The Four Focuses

This section describes the Four Focuses – those educational content areas that provide the backbone for an industry-ready education. In the next section, we put the contractors' wants together with these Four Focuses to derive some key lessons for education and the industry.

Vocabulary

Vocabulary is the set of proper labels and definitions of concepts. Words and actions are all we have to interact with others; a good industry-ready vocabulary is critical for the new graduate's credibility and effectiveness. Precise vocabulary lends gravitas to the speaker. It gives her or him perceived depth of understanding. Someone with a functioning industry vocabulary can communicate with fewer misunderstandings in less time. Everyone wins.

This vocabulary focus – the use of the right word or phrase – comes only from understanding the accurate meaning of that word or phrase. To learn this is a remedial exercise – less exciting than technology improvements or risk strategies, but no less important. Words are a part of the language, and without language, we cannot think as well nor communicate efficiently. Using precise vocabulary is less risky. Proper words form the basis of accurate conceptual frameworks, which leads to our next focus.

Conceptual Frameworks

Having a "mental template" helps the construction professional filter and interpret what she/he sees and hears. That same template can help them explain their idea. The first conceptual framework we suggest is that success in construction is always predicated on:

- Safety
- Quality
- Cost
- Schedule

These have obvious value. Providing these four outcomes to Construction Service Buyers makes for better business relationships. Relating every issue to the possible effects on Safety-Quality-Cost-Schedule is a great "best practice."

Sometimes, these mental templates can be visually shown. Stevens's book, *Managing a Construction Firm on Just 24 Hours a Day* is populated with graphical representations of many critical concepts for contractors to understand. Dynamic models such as Risk-Reward, Supply-Demand, Overhead Cost Application, or Target Marketing. We purposely inserted these graphics as a basis for quickly communicating important considerations when making a decision. As everyone has heard at least once, "a picture is worth 1,000 words." We count more than 40 different construction contracting concepts that can be graphically displayed and labeled. It is important to state that we believe most construction professionals are visually attuned.

We have used this book and its graphics as the basis for the learning and teaching of university students. As part of the final assessment, the Instructor gives a list of concepts and asks students to sketch each, labeling all parts and giving a description of its meaning and application to construction contracting. As you might guess, it is the rare student who scores a 95% or above. It is so few that the Instructor can name each of those students. Affirming our belief in this graphical technique, several previous students have communicated that this approach has been helpful in their career. Many times, they use graphics to explain their thoughts better. From this, construction graduates might be well prepared to sketch graphically more often their directions to field craft, operator, and management leaving less ambiguity about their instructions. Most times, it is directed at subordinates, sometimes peers, and on a few occasions, superiors. See Figure 1 for example.

Figure 1 shows a critical concept for students to understand in construction management. Whether a main contractor or subcontractor, each firm has to plan, communicate, execute, measure and adjust each of the four activities shown. The

Figure 1. The business cycle of construction contractors
Source: Matt Stevens Ph.D.

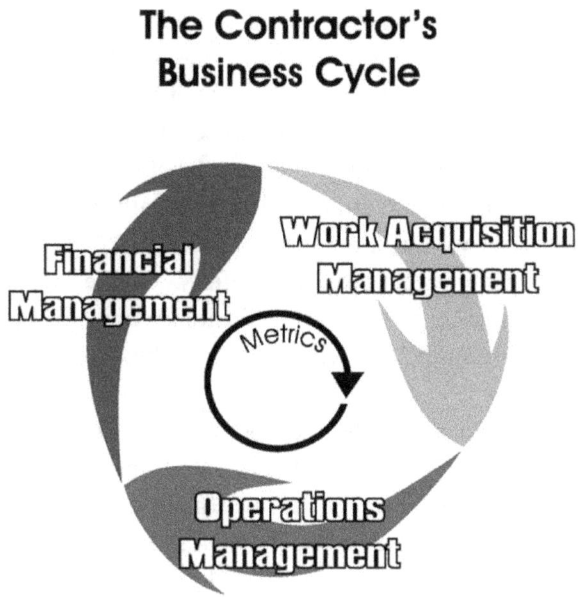

cycle of the business of construction contracting helps students clarify their thinking and at times, filter and organize an overwhelming amount of information such as a case study.

Several points might be made in the presentation of this such as, "The speed of the cycle determines your competitive edge" and "where it slows down typical is in the handoff of information between each step." This will give students some ideas about the depth and applicability of this framework to the "real world". After many of these types of graphics presented and explained, students are ready to think about and analyse project situations and company crisis narratives.

As an example of learning outcomes, one of our graduates impressed her employer with her crisp understanding including accompanying visual explanations of 1) Margin versus Markup 2) Project Return on Investment 3) Construction Risk versus Reward 4) Strategic Planning among other critical concepts. Certainly, middle managers received a refresher on these applicable ideas at the same time.

We have had several of our young graduates participate in a strategic planning session(s) with senior staff. Why? They know the fundamentals of the construction industry and emerging changes. Their up-to-date knowledge helps keep "guardrails" on wandering discussions that reminisce about the past. In all, they assisted leaders to keep the conversation relevant and moving forward.

Iterative Problem Solving

"The first answer is the wrong answer," as one of our mentors says. That is, those who hurry to conclusions as if they are on a television game show are often wrong. Even if correct, they many times have not thought of their answer thoroughly so, it will not stand the test of time. Details matter in construction contracting.

Careful and often slower thinking facilitates better answers, decisions, and qualitative judgment. Any construction challenge usually possesses a complex set of dynamics. Only when we question, and re-question do we give ourselves time to collect and analyze more information carefully. As the last step, interjecting subjective considerations helps, so our conclusion(s) are holistically strong. For the Instructor, using case studies and challenging problems help show the value of iterative thinking

We present several scenarios in the classroom to emphasize this. One is a 4-mile route and an earthmover's speed to show the need to think and compute first before answering. Another is a project return on investment calculation challenge. Still, another is a quiz about the cost of rebar per ton and concrete per yard, together and separately. One of our favorites is a bridge-building exercise – a bidding and building competition. For case studies, we have used our books' cases and other's construction-centric ones from publishers such as Harvard Business Review. During these exercises, our instruction to students includes that famous NASA saying, "work the problem."

The emergence of intercollegiate competitions in areas such as Design-Build and Heavy Civil have much the same effect. Each school's team is given a set of plans, specifications, and scenario with which to determine an approach. In a sequestered setting, they are left to think through the challenge with only each other. From our understanding, iterations of thought are in the dozens. This is realistic to problem-solving in construction. The judging panels are industry professionals who deliver the final iteration.

Iterative thinking is often slow but, the final product will stand the test of time. It is a habit to instill and nurture. The resulting thoughtfulness and detail orientation are often high. As proof, semester-long project-type learning is the norm in many elite university programs, construction or not.

Writing

Writing leads to thinking and speaking. If writing is improved then, we have seen an improvement in the other two. Iterative writing with feedback from a knowledgeable Instructor provides clarity. Well-expressed thinking is not as evident in the construction industry as contractors wish. This means to us that graduates who have clarity and organization in their writing are more valuable than other applicants.

One compelling reason for focusing on writing, liability risk is high for construction contractors. Contemporaneous, clear and complete writing protects contractors. Quality writing is only possible if the previous three focuses are ingrained. That is, vocabulary identifies, conceptual frameworks organize, and iteration produces complete writing.

Writers must research to generate impactful content. Whether it is delay history or re-mobilization costs on a project, these facts have to be discovered and vetted for truth before they are committed to paper then sent to the client. In the classroom, this research sensitivity is started by lectures citing data sources and later with challenging questions ("Can you find this answer?") on students' submittals.

Organized and articulate content earns creditability for the writer (and speaker). Words, concepts and their expression in a logical and straightforward way deliver information that can convince, persuade, instruct or protect. Again, we observe that the quality of thinking and speaking follow.

Strong writing is centered on content, not formatting or rules, although these are also important. Errors are always a concern. To keep mistakes low in number, an editor – (or now, software) – can help. The focus on the content of the one's writing is critical. Content is many things, such as a fact, argument, insight, connection, or explanation. When the construction professional who is trained in writing is immersed in the facts of a complex situation, clarity emerges, and a reasoned path forward can be created whether in discussions or on paper.

Teaching in the way we recommend can be a challenge, but in some ways, it is simpler. At Western Sydney University where Stevens teaches over 30% of students speak a language other than English at home. Additionally, over 20% are born outside of Australia, and approximately 40% are first in their family to attend university. This means that extensive writing may seem challenging, but we seek only to make people competent technical writers. Almost all are proficient in English due to the university's requirements so, it is just a matter of continuing coaching and encouragement.

Everyone can learn to write competently. In our experience, this is true. We have used the example of writing a loved one's eulogy. Any person who has lost someone dear can write a compelling message about the departed. The focus is high, and the deceased's essence is clear to the writer. The eulogist has one chance to deliver their thoughts. Many times, there is should be no holding back of content. Due to the personal nature of such as task, the eulogy will be rewritten multiple times until it states what is intended.

Putting It All Together: The Formative Process

The formative process delivers those features that contractors want most in graduates: thoroughness, independence, steadiness, and integrity. It does this through iteration. The process is a continuous process throughout a semester. This is a long time to think and rethink thoroughly about the subject matter at hand. We might consider it a cycle. With this in mind, there are sub-cycles - from the introduction of a concept to the student's mastery and application of the same concept. A key step in-between is the instructor's check on understanding or misunderstanding.

The Formative Teaching, Learning and Assessment's goal is to constantly monitor the learning progress of the student and adjust the teacher's delivery for real-time improvement. Questions such as, "How could this class be taught better" to the entire class helps the instructor monitor the pulse of student mood and learning. We have found that there are fewer surprises in the student evaluations afterward.

Other questions to the entire class such as, "Are there any blind spots that I have not covered so you can write a very good paper?" and "If you were going to teach this class, what else would you cover?".

Of course, the formative process of learning may be longer. A Ph.D. dissertation is an example of a multi-semester one. It involves a disciplined process while learning new concepts and mastering them shown through writing a lengthy document with others monitoring for quality. The Ph.D. committee typically checks the document several times a year for accuracy and completeness of thinking. In contrast, the summative process is one where the student learns and then is assessed at the end of an educational unit. Commonly, this occurs several times a semester with quizzes and tests. Many times, without an active connection to previous topics.

Teaching in the 4mat Process

We will use the 4mat model as our basis of formative teaching, learning, and assessment. The 4mat process is a sequential teaching model created by Bernice McCarthy in 1987. See Figure 2. It is an orderly approach to learning which follows a human's nature. The graphic shared can be used as a quick checklist to create teaching plans for construction management students. We will use the graphic to help describe the 4mat process.

Quadrant 1 is primarily focused on building learner interest; we recommend creating an experience that engages people in learning and gets them excited to learn more.

This experience should draw on the prior knowledge and experience of the group and tap into their perceptions. So, it is an engaging approach that may mean specific questions will be asked and interesting stories told.

Figure 2. The 4mat teaching model
Source: Dr. Bernice McCarthy, About Learning.

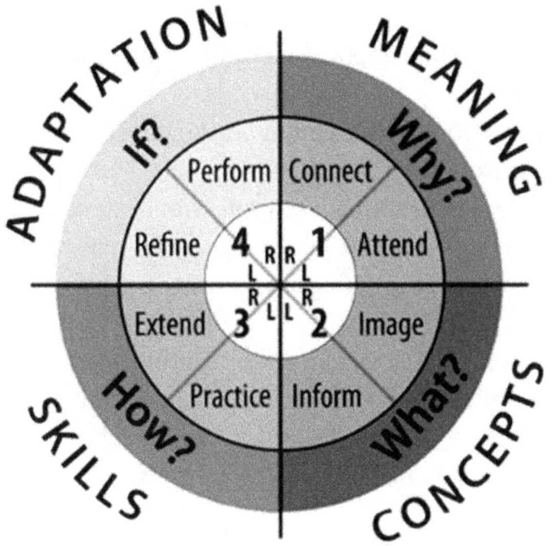

Quadrant 1 is about the classes' experiences including personal perceptions, and values while answering the question, "Why should someone learn this content?". Interestingly, students, these days are questioning the value of college in new ways. This is a timely approach to make sure value is delivered by the teacher.

Quadrant 2's goal is to communicate the significant ideas and concepts comprising the material and helping to clarify what is essential for people to know.

This instruction phase is where students are taught key information you are trying to impart. Quadrant 2 is about understanding the key principles, ideas, frameworks, or concepts. In the world of construction management education, a graphic(s) should be used as part of the presentation to increase the "stickiness" of the content.

Additionally, it has been a helpful activity when a student creates a graphic to illustrate the concept or idea that they have learned. If they can label and explain the model in great detail that is a sign of deeper than average learning.

Quadrant 3 is where the student is more active and begins to take over. They must master the content and become skillful in the use of learning. This is when connections to other concepts and principles are made. These are significant insights that show a valuable skill to be used later in their professional life.

After mastering the content then students can work to extend learning into their own lives and transfer learning skills into real-life applications through hands-on experimentation and problem-solving.

Quadrant 3 is about learner mastery (and not performance against other learners), so it is important to provide a wide range of practice exercises both individually and in groups. It should be expected that the student sees connections to other content. Additionally, they may see the application of the content to current events or their individual experiences

Quadrant 4 is where people create new understanding through personal evaluation and or extension of learning.

If the student is to be a true emerging professional should know the topic well. To demonstrate mastery, it must be put to use in the real world and adapted in a personally meaningful way. The writing assessment portion of our teaching approach will show that to the teacher.

For the instructor, personal stories are best, since the instructor can add rich detail. If none exists, then application by the instructor to a current event shows relevance to the construction industry and makes the concept real. Additionally, asking the student for what applications makes sense to them.

Formative Teaching in Action

We use the 4mat process as a template for lesson planning for each class. Thoughts and connections are prompted by each step of the model. In Table 1, we answer the question posed by quadrant in for each classes content.

For example, summarized in Table 1 is a brief lesson document on Pre-construction planning to demonstrate.

This pre-construction content is the needed content to help students as they write their assignment. The instructor, upon grading, will review for precise vocabulary, conceptual frameworks as well as a careful process of explaining their approach.

Student Writing Assignment Process

We have used the formative approach extensively in the classroom. One of us who we will refer to as the Instructor, structured and implemented the following Formative process. As a specific example, the Instructor taught a specialty contracting course and required each student to select a trade. From that, they wrote a company operating plan. This plan included the long-term business strategy along with how the company processes of acquiring work, constructing projects, tracking the results and human resource management. Weekly lectures' containing content followed the same sequential order.

Each student would submit their expanding writing project every three weeks with a first version ungraded but reviewed and comments given by the instructor.

Table 1. Sample lesson planning document for pre-construction content

Pre-Construction Planning Lesson Plan		
Question	**Information to Share**	**Points to Make in Class**
Why?	-The most valuable single planning event in the entire project life. -If you can lead it well, you will have high professional value	-One chance to do it well, otherwise opportunity is lost forever. -Getting every person on your project team excited and focused has great benefits throughout the project. -Share graphic
What?	-Acquire extensive information before formally planning with others	-Project estimate -People availability -Equipment needed -Phasing of project work packages -Cash flow projection -Risk assessment -Use calculations to prove
How?	-The steps are several (go into detail of each step)	-Make sure that everyone involved has time to study information -Gather perfect information -Start quickly after winning project – people's memory is fading. -Share observations
If?	-Compare a small and large project – is the process different?	-What are the areas that differ? Why? -Give experiences
	-What are common errors made?	-No senior executive attends every meeting -Starts after a long delay – memories are faded -No action plan created or followed-up -Give examples
	-How long into the project is the period considered pre-construction planning?	-A pre-construction plan is not a plan unless it is written and executed -People may not tell you about the barriers they face the first weeks of a project. Supervisors have to ask and then knock down barriers. -Tell stories

Source: Matt Stevens Ph.D.

This "free" submittal gives the students an idea of what is expected. This unmarked first submission has compelling reasons behind it. Here are a few:

1. No one is penalized for misunderstanding the Instructor's requirements. This appears to help students calmly follow directions and not overthink the assignment.
2. To see who is more serious about the course. Typically, 25% of class members will not submit. This is not indicative. Some students may have had a bad week and put off the stress of a submission. However, it prompts a discussion about ignoring an opportunity between students and Instructor.

3. The Instructor sets expectations. All papers have comments asking questions about clarity, content, or connections with other ideas. For the best students, the Instructor may ask questions to extend their thinking. For those unmotivated, the comment, "this is not a passing paper" has prompted more intense work.
4. Grammar check for those who may need help. This early assessment of language understanding is helpful. It prompts the Instructor to remind some students to adhere to basic rules. This is important in this age of more international students. Again, content is primary in strong writing.

Through iteration, students develop the vocabulary, conceptual frameworks, and writing skills necessary to succeed in construction. It becomes their default. Each submittal contains instructor's comments which must be answered with corrected or updated text. With this iterative and careful review of student work while mentoring their learning process, comes the thoroughness, independence, and steadiness that contractors seek. Integrity arises at least partly out of the combination of these processes (vocabulary, conceptual frameworks, thoroughness, independence, and steadiness). The instructor's insistence of timely work perfected over four months discourages "excusitis" that some students may default to. As most know, the industry dynamics reward construction firms when tasks are "done, done" on-time – if so, safety improves, quality is higher, and productivity exceeds competitors. So, an Instructor's insistence on completeness and timeliness of work transforms students' thinking to align more with the realities of our industry.

There are other ways that the transformation of thinking can occur. We offer the example of internships. These compensated work programs have been a "tipping point" for many students. They are immersed in the culture and up-tempo atmosphere of the job site along with a good-natured indoctrination from experienced professionals. Afterward, they return to school with new-found knowledge and accompanying experience more aware than ever of the importance of what is being taught inside the classroom. A confident, prepared employee understands what she/he is seeing, asks questions when needed, knows the difficult moral and ethical issues, can effectively argue for a safety "stand-down" or against bid shopping.

In practice, if an undergraduate construction program followed a complete formative process, its beginning courses would emphasize vocabulary, foundational concepts, exercises, and written assignments for areas such as safety, quality, cost, and schedule. Later, classes expand to cover components of each such as ergonomics (safety and long-term health), eliminating rework and workarounds (quality and planning), labor productivity (cost and quality) and scheduling (planning and risk management) in its end of program courses. Each class would have a healthy dose of continuous writing. Many times, a graphic(s) would be required to be used to explain salient points further. Her/his writing would be electronically checked for originality, and on-time submittals are gauging conscientiousness. A couple of summer internships consolidate knowledge and perspective. One or more intercollegiate team competitions bring understanding into sharp focus. During this journey, students are expected to create connections or insights into disparate concepts, i.e., how does safety affect productivity and vice versa?

As an everyday example, if a person is passionate about Virtual Design and Construction (VDC) – mastering the four focuses is a pathway we have seen work. At the end of their formal education including outside classroom activities, writing about VDC's challenges or new developments clarifies her or his thinking about such things. To wit, they know what they think and why they think it. This is an asset in any serious discussion. Also, clarity of thought begets insights. Insights may be breakthroughs that others have never realized. This transformation of thinking is core to what we think construction education is about.

Quite frankly, we are thrilled when students "find their voice." Professors and publication editors confirm that this is a real phenomenon. The students enjoy it the most. They can confidently use a powerful tool with which to transform other's thinking. In our experience, it takes much work but is well worth it.

As an aside, if we were to outline an online course for the construction industry, we would use these four focuses as our basis. The Instructor has experience with facilitating these kinds of classes. We know that there is an impersonal nature to e-courses, so the Instructor's personality and artfulness is not readily evident to the learner. However, instruction and assessment for vocabulary, conceptual frameworks, solving complex scenarios and writing progression dovetail nicely into distance learning.

Overall, whether in-classroom, online or a blended approach, the described structured process of inputs and outputs in teaching, learning and assessment appears to produce better outcomes.

Common Criticisms of Our Recommendations

We know others will criticize some recommendations. This is to be expected and is a healthy part of the discussion we are seeking to prompt.

- **The Focus is on the Science of Teaching, Not the Art**: Many believe teaching is more art than science. We don't disagree. This white paper may reduce teaching to a checklist of actions (science) with no regard to the soft skills (art) required to generate student motivation and understanding in the classroom.
- **Instructor Time:** This is a factor; however, with the development of electronic means of submission, reviewing, commenting, and plagiarism checking, the

assessment work is easier than before. However, the Instructor can attest to more time needed than a summative assessment. We think the time is worth the investment.

- **Program Time:** Some may say that we are adding time to an already overwhelming set of requirements for a construction program to meet. We hope not. The focuses and formative approach do not add to existing curriculum nor time needed per class. With our suggestion of a writing component to each of the final program courses, we may be adding to the instructor's assessment burden. For that, we are guilty.

CONCLUSION

This Chapter prompts discussion about improving the quality of contractor desired skills of college graduates who enter the construction industry. It is not exhaustive nor does it pretend to contain best practices, only what we have found that it is effective for us. We do not suggest that this is meant for those who will work for firms engaged in such sectors as project design, owner's representation, or project inspection. Since the focus of our work has been exclusive with construction contractors, we hope you find these opinions credible.

Ironically, we have learned some of this from others - the great teachers who work inside and outside of construction – but we are prompting those same teachers (and others) to consider a slightly different approach.

It is clear to the industry that today there is a shortage of qualified managers. This signals to us that a large opportunity exists for any construction program. That is if they can graduate students with an understanding of needed vocabulary and conceptual frameworks combined with iterative problem solving and writing skills. Construction executives talk to each other and word travels fast. Increases in career day attendance and interview schedule slots are signs of more valuable graduates.

We hope that the framework described in this Chapter results in a more intense discussion and a closer relationship between construction professionals and the educators on which they rely. Whatever the subject is, we think the four focuses and Formative Approach help. As we stated, they do not affect content. In total, we suggest that there may be strong learning outcomes if a consistent focus and delivery method are maintained throughout a student's education.

Formative teaching, learning, and assessment is an effective approach in our experience. There is much research to confirm its value. This process facilitates many positive outcomes such as better 1) vocabulary capture 2) mental templates, 3) iterative problem solving, and 4) writing skills. We assert more continuous writing assignments will improve these four even more. Outside the classroom, internships and intercollegiate competitions have a proven transformation effect and are a given.

Each year, industry professionals witness new requirements and realities. Construction education helps address industry change with teaching updated subjects and introducing new ones. There is no predicting the industry's future path, but in many respects, it will not follow the past. Therefore, we believe that preparing for more change possibilities with up-to-date and deeper thinking is the wisest choice. From our perspective, we are seeking to make students more thoughtful and self-reliant while ethically sensitive is the answer. From these strengths, we know that producing low maintenance professionals who can add value to the construction industry is more certain.

REFERENCES

Bloom, B. S. (1968). *Learning for mastery.* Los Angeles, CA: University of California Press.

McCarthy, B. (1990). Using the 4MAT System to Bring Learning Styles to Schools. *Educational Leadership*, *48*(2), 31-37.

Stevens, M. (2012). *The Construction MBA*. McGraw-Hill Education.

Chapter 3
Group Dynamics in Higher Education:
Impacts of Gender Inclusiveness and Selection Interventions on Collaborative Learning

Mehrdad Arashpour
Monash University, Australia

Julia Lamborn
Monash University, Australia

Parisa Farzanehfar
University of Melbourne, Australia

ABSTRACT

Although group work has been proven to be an effective method for enhancing active learning in the higher education, optimum planning is crucial for successful implementation. A deep understanding of teamwork dynamics and creation of inclusive environments helps groups to demonstrate their optimum performance and output. On this basis, the current research focuses on the important challenge of gender inclusiveness and required teacher interventions to encourage that. Towards this aim, three research hypotheses are developed and tested using student performance data in a series of individual, group, and hybrid assessment. Findings show the significantly different performance of female and male students in group activities. It is also found that instructor interventions to form gender-inclusive groups significantly improve group performance and output. This works contributes to the higher education literature by exploring dynamics of collaborative learning and interfaces with gender inclusiveness. Educators can utilize the findings to better design and implement team activities.

DOI: 10.4018/978-1-5225-8452-0.ch003

INTRODUCTION

Within the higher education context, collective learning and group work are found effective for developing capabilities, qualities and skills in students that are considered as important graduate attributes (Denson & Zhang 2010; Caple & Bogle 2013). Previous research has shown the interdependence between group learning and graduate attributes related to employability and resilience (Arashpour, Sagoo et al. 2015). These justify the investment of higher education providers in developing modern learning infrastructure that facilitates 'learning in the round'. Interactive lecture spaces are often equipped with cutting-edge technology to enable peer presentation and group collaboration (Asok, Abirami et al. 2017; Stache, Barry et al. 2017). Sometimes design of learning spaces can be self-configured by students to better suit collaborative activities (see Figure 1).

Although modern and interactive spaces provide the required infrastructure for collaborative work, the role of instructors in facilitating 'learning in the round' is still crucial (Arashpour & Aranda-Mena 2017; Khuzwayo 2018). Educators should encourage student groups to work efficiently on collaborative assignments with the aim of developing decision-making skills, communication and critical thinking (Schaber, McGee et al. 2015; Debuse and Lawley 2016). Towards this aim, important factors in designing group activities include but are not limited to optimal group size (Scager, Boonstra et al. 2016; Francis, Henderson et al. 2018), member selection methods (Joshua & Mariajose 2013; Poelmans & Wessa 2015), proper scheduling

Figure 1. Collaborative learning spaces for facilitating group works

of teamwork (Mangum, Johnson et al. 2017; Arashpour, Miletic et al. 2018), and setting challenging tasks (Adwan 2016; Zschocke, Wosnitza et al. 2016).

The aforementioned factors can significantly influence dynamics of teamwork; however, they do not necessarily optimise member interdependence, involvement, and fair division of labour (Kalfa & Taksa 2017; Maina, Oboko et al. 2017). One variable that has been largely overlooked by the higher education literature is the influence of gender-inclusiveness (Beylefeld & Le Roux 2015; Smith & Kellogg 2015). This variable can provide essential intellectual and social resources to groups and encourage both female and male members to develop a sense of responsibility towards the team success (Magogwe, Ntereke et al. 2015; Arashpour, Bai et al. 2018).

In order to bridge this gap, the current chapter aims to objectively analyse the impact of gender-inclusiveness on group performance and output. Achieving gender-inclusiveness is a nontrivial challenge within the collaborative learning context and requires educator interventions in group formation and member selection (Lee, Kim et al. 2017). The dynamics of such interventions will also be analysed in this chapter. Towards the objectives of the current research, academic performance data of 360 students in different group and individual learning activities were collected. Quantitative analysis of data formed the basis for testing three research hypotheses around collaborative learning.

The chapter is structured in the following format. The next Background section provides an overview of most common approaches to group composition and member assignment. The section also looks at gender-specific motivational beliefs in the context of higher education. In Section 3, the adopted research method and framework are described, followed by development of three research hypotheses. Analyses and results are presented in Section 4 and finally conclusions are drawn in Section 5. Opportunities for future pedagogical research on collaborative learning are identified at the end.

BACKGROUND

Tertiary education literature has shown the importance of student group work and the link to employability, future career success and productivity (Huff 2014; Ali, Tahir et al. 2017). The main objectives of group work in classrooms are to develop skills including communication (Goodman, Knight et al. 2014), cooperation (Arashpour, Too et al. 2017), collaboration (Maqbool, 2016), and compromise (Beylefeld & Le Roux, 2015). Considering the importance of group work and its objectives, tertiary educators need to make every effort to understand dynamics of group activity in order to enhance student experience. The following sections provide background information from the literature on two important variables of member assignment and gender differences that have major influence on group dynamics.

Group Member Assignment

There are several methods by which students can be assigned to working groups including self-selection, random assignment, hybrid methods, and purposeful assignment. Self-selected and instructor-selected groups are both very common in the tertiary education context (Ficano 2018). In self-selected groups, students can freely select the members they want to work with. Previous research has shown the superior performance of self-selected groups in terms of communication between members (Elahi, 2017), confidence in the collective competence (Goth, Bergsli et al. 2017), and enthusiasm to work together (Hessler & Henderson, 2013). Self-selected groups, however, can face problems with regard to time management (Moore & Hampton, 2015), completing work of other members (Van Leeuwen, 2016), and failure to follow commitments (Wang & Vollstedt, 2014).

Instructor-selected groups are usually formed by random assignment of students to different groups (Hampton & El-Mallakh, 2017). Furthermore, group composition can be determined by adopting a hybrid approach by which students first have the opportunity to form groups themselves and then instructors assign remaining students to existing or new groups (Wang, Tsiligiris et al. 2018). Another approach to form instructor-selected groups is to purposefully assign students based on some assignment-related criteria (Khosa and Volet 2013). Instructor-selected groups minimise the problems of free riding (Ali, Tahir et al. 2018), and boredom caused by working with same circles of friends (Nasirian, Arashpour et al. 2018). Instructor-selected groups, however, are not usually favoured by students (Herr, 2013), and members are less likely to work together in future group activities (Thrall, Coverdale et al. 2016). Reviewing the tertiary education literature suggests that member assignment methods have significant impacts on group dynamics and should be chosen based on course and cohort characteristics (Smith & Kellogg, 2015).

Gender-Specific Motivational Beliefs in the Higher Education Context

Contemporary tertiary education reviews reveal that gender differences continue to exist (Woolard, 2018). Previous research has focused on the root causes of gender differences in learning and performance, resulting in identification of important factors such as learning goal orientation (Skelley, 2014), task value (Harding, 2018), and self-efficacy (Hyldegård, Hertzum et al. 2015). Learning goal orientation (LGO) undermines both motivation and achievement for male and female students

(Lambert, Carter et al. 2014). A solution to maximise LGO is to frequently emphasise on learning goals in different stages of learning (Howe, Sheu et al. 2018). Another integral factor in learning is task value recognition by students. This plays an important role in student performance and achievement-related choices (Guilbaud & Whitney, 2018). Previous research proves that male and female students are more inclined to complete learning activities if the value for future is understood (Arashpour, Wakefield et al., 2018). Understandably, students' perception of task value is impacted by gender differences and male students demand more instructor effort to recognise task values (Odeh, McKenna et al., 2017).

Self-efficacy beliefs can explain many gender differences in learning and academic performance (Streiner, Besterfield-Sacre et al. 2016). Self-efficacy determines whether or not female and male students believe in their learning capability (Scott, Mistry et al. 2014). Theorists contend that past research related to gender differences in academic learning has frequently centred on individual performance and has often neglected group activities and their dynamics (Zitzelsberger, Campbell et al. 2015). The objective and contribution of the current study is to bridge the existing gap by focusing on group dynamics in higher education under influence of important variables such as gender differences and member selection.

RESEARCH METHOD

Framework

Based on the theory and research on group activities, a research framework for analysing dynamics of group work was developed. Different higher education courses were selected in which there were both group assessments and individual assessments including projects and final exams. Collected student performance data were analysed to test the three developed research hypotheses. Altogether, 360 students comprising 207 males and 153 females participated in this study.

Table 1. Median and interquartile ranges of assignment marks for the study population

N= 360 Female 42.5% Male 57.5%	Marks Median (IQR 25%-75%)
Individual Project	**85** (75.6-92.5)
Group Assignment	**80** (76-88)
Final Exam	**80** (68.5-86)
Total Score	**81.5** (72.2-87)

The students were in years two to four of their undergraduate studies in two public Australian universities. Table 1 shows the median and interquartile ranges (IQR) of assignment marks for the study population.

Hypothesis Development

Previous research has proved the importance of motivational beliefs on academic performance of tertiary students (Abokhodair, 2015; Lee, Kim et al. 2017). Among the others, self-efficacy, recognition of task value, and learning goal orientation are most influential motivational beliefs (Hu, 2015; Crawford & Jenkins, 2017). On this basis, pedagogical interventions to enhance motivational beliefs can be beneficial (Fuji, Tanigawa et al. 1996; Meschitti, 2018). However, such interventions have not been customised to reflect unique characteristics of learning in female and male students (Botterill & Harpe, 2010; Noguera, Guerrero-Roldán et al. 2018). To address this nontrivial challenge, the first step is to measure how different female and male students perform in the higher education domain and particularly within the collaborative learning context. Towards this aim, the validity of the research hypothesis needs to be tested,

Hypothesis 1 (H1): Performances of female and male students in group activities are not significantly different.

One of the main influencers on the group performance is the method of group formation and member selection (Ng & Hanewald, 2009; Coccoli, Stanganelli et al. 2011). As discussed in the background section, groups can be student-selected, instructor-selected or a hybrid of the two (Rae, Gee et al. 2009; Mendo-Lázaro, León-del-Barco et al. 2018). When students are allowed to self-select their group mates, the tendency is to choose close friends and then additional members are picked from second and third circles of friendship (Brooks and Moriarty, 2006; Adya, Temple et al. 2015). Furthermore, instructor-selected groups are often formed by random assignments of students, which is less time consuming for instructors (Kampe & Oppliger, 2011; Clapp, 2017). None of the aforementioned group formation methods consider the individual performance of members in previous assignments and the potential influence on group performance as a collective. On this basis, the second research hypothesis can be advanced as,

Hypothesis 2 (H2): Group performance in assignments is not correlated on the performance of associated members in individual assignments.

The final null hypothesis of this research integrates H1 and H2. It was discussed in the background section that instructor intervention in group formations can accelerate the process, which is particularly important when there are early assignments within the teaching period (Lalopa, Jacobs et al. 1999; Yuretich & Kanner, 2015). Sometimes instructor interventions aim a purposeful selection of group members based on assignment-related criteria such as required skill sets (Feldmann, 2006; Han, Liau-Hing et al. 2016). Although instructor interventions to from groups by considering assignment-related criteria have been investigated excessively (Keevers, Lefoe et al. 2014; Yannibelli & Amandi, 2017), only few studies have evaluated instructor interventions to create gender balance within groups. This may suggest that instructor interventions for creating gender balance within groups do not create significant impact on group performance due to independency of group output and members' gender. On this basis, the last hypothesis in this research is developed as,

Hypothesis 3 (H3): Instructor interventions to form gender-inclusive groups do not significantly impact group performance and output.

ANALYSIS AND RESULTS

As discussed in the previous section, individual and group marks of 360 students were analysed across a variety of assignments and exams. The first null hypothesis proposes that performance of female and male students in group activities are not significantly different. In the first step to test the hypothesis, descriptive statistics including median and interquartile ranges (25%- 75%) were used to summarise the collected data (see Table 2). As can be seen, for the studied population sample, the median (50th percentile) and IQR of achieved marks has been higher for female students in both individual and group assignments.

Although descriptive statistics are helpful in measuring central tendency and variability of data, inferential statistics are required for an in-depth interpretation of

Table 2. Effect of gender difference on assignment outcomes

	Female Median (IQR)	Male Median (IQR)	Effect Size	P-value (t-test)
Individual Project	**90** (77.5-92.5)	**82.5** (75-92.5)	0.7	0.1
Group Assignment	**88** (80-92)	**80** (72-84)	0.8	0.05*
Final Exam	**82** (70-90)	**77** (64-84)	0.5	0.1
Total mark	**85** (73.5-89.5)	**78** (70-85)	0.75	0.02*

Figure 2. Relationship between individual and group performance: (a) Influence of individual performance on group score, (b) Influence of group performance on final individual score

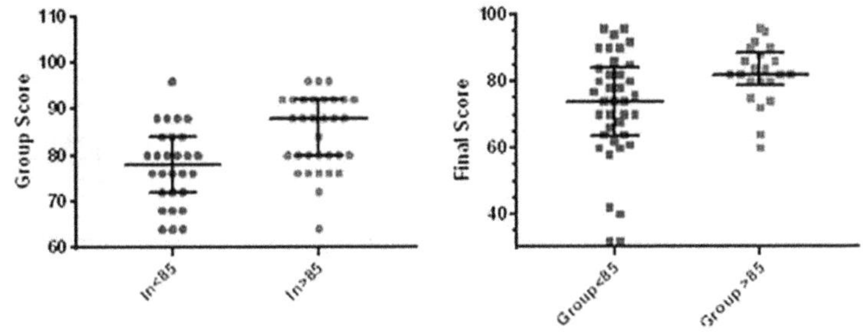

trends (Arashpour, Kamat et al. 2018). On this basis, t-test was deployed to measure the significance of difference between achieved marks by female and male students. Furthermore, Cohen's effect size was calculated to translate the difference between the two groups into standard deviation units.

As Table 2 shows, the p-value derived from t-statistics indicates the significant difference between performance of female and male students in group activities. More importantly, the effect size of 0.8 suggests that performance of the two groups differ by 0.8 standard deviations. Both descriptive and inferential statistics result in rejection of H1 and confirm that performance of female and male students in group activities are significantly different.

The second null hypothesis (H2) proposes that collective group performance is not correlated with the performance of associated members in individual assignment. In order to test H2, box plots of data assisted the mapping of the individual performances in assignments against group performances. As can be seen in Figure 2(a), groups comprising higher achieving members (individual marks equal or above 85%) were able to perform better in collaborative assignments by achieving the median mark of 89%.

In the second step to test the correlation between group performance and exam final marks, the two were mapped against each other (See Figure 2(b)). Understandably, individuals associated with high achieving groups (with marks equal or greater than 85%) managed to achieve higher final exam marks. Therefore, the results of the two analyses dos not support H2 to stand. This finding is in line with that of Harding (2018) and confirms that students of a feather tend to flock together!

To recap the investigation to this point, testing the first null hypothesis revealed the existing non-similarity in the performance of female and male students in the context of collaborative learning. Furthermore, hypothesis testing for H2, proved that group performance is correlated with the performance of associated members

Figure 3. Importance of gender-inclusiveness for assignment performance: (a) Correlation of gender and individual performance, (b) Correlation of gender and group performance

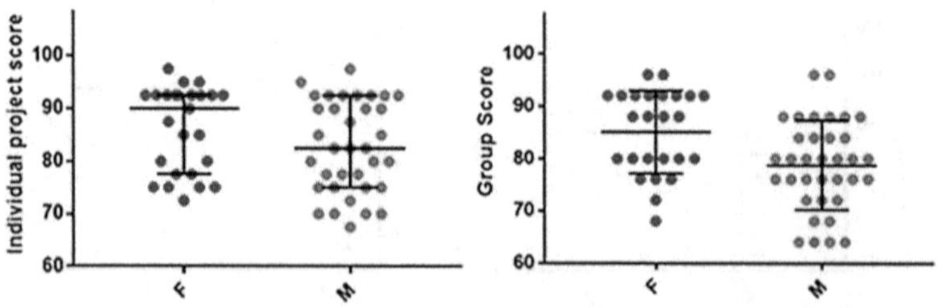

in individual assignments. Considering the higher performance of female students, it can be assumed that forming all-female groups will potentially result in superior group outcomes (Chen, Wu et al. 2006; Matveev & Milter 2010). In order to integrate H1 and H2, the third null hypothesis focuses on instructor interventions for composing gender-inclusive groups and impacts on performance and output. Such interventions may also enhance student experiences in the process of collective learning and facilitate fairer group performance assessment (Serçe, Nur Alpaslan et al. 2010; Brewer, Mitchell et al. 2015; Bui, Baruch et al. 2016).

In order to test the third null hypothesis, the individual performances of students were mapped against their gender. As Figure 3(a) illustrates, there is a significant difference between the two group performances. In the second step, performances of students in group assignments were mapped against their gender (see Figure 3(b)). As can be seen, both median of achieved marks and interquartile ranges for female students are higher than male counterparts, with effect size of approximately one standard deviation.

Results of the above analysis show that gender has a significant impact on individual and group performances. Considering the H2 hypothesis testing and the proved correlation between group output and individual performance of associated members, it is concluded that instructor interventions to maximize gender-inclusiveness in group compositions are necessary. Otherwise, there will be a significant gap between performance and output of all-male and all-female groups. As a result, the third null hypothesis (H3) cannot be supported.

CONCLUSION

The optimisation of collaborative learning and student experience in group work has long been the focus of research in higher education literature (Chapman, Meuter et al. 2006; Angelva, Tepsa et al. 2017). However, only few studies have adopted a holistic approach that considers important performance variables such as gender inclusiveness and selection interventions in collaborative learning (Awatramani and Rover 2015, Arashpour, Bai et al. 2017). In order to bridge this gap, the current research analysed individual and group performances of a large population sample of tertiary students to shed light on dynamics of group work in higher education. Results of analysis were used to test three research hypotheses around performance of different genders in group work, relationships between individual performance of associated members and group outcomes, and finally the role of instructor interventions to form gender-inclusive teams.

Using both descriptive and inferential statistics, it was found that female students generally outperform their male counterparts in group assignments. This superiority creates a tendency to form all-female groups with the aim of achieving higher marks in group exercises. Therefore, instructors' interventions will be required to maximise gender balance in all groups. Considering the diversity of motivational beliefs and self-regulation in male and female cohorts, such interventions will optimise learning goal orientation, task value, and self-efficacy that are key players in group output and performance.

FUTURE RESEARCH DIRECTIONS

There is need for comprehensive research on the impact of instructor interventions on students' experiences within team-based learning settings. This research is particularly important due to tendency of higher education providers to link the instructors' performance to course experience surveys. Instructors will be less likely to intervene, if an adverse effect on student experience and therefore instructor performance is found. Future research should also adopt a longitudinal approach to monitor the impact of gender-inclusiveness on work-readiness of graduates and professional success of alumni.

REFERENCES

Abokhodair, E. S. (2015). Action learning as an approach for developing leaders, solving problems, and building successful teams: A case study of princess nora bint abdul-rahman university. In Community Engagement in Higher Education: Policy Reforms and Practice. Sense Publishers. doi:10.1007/978-94-6300-007-9_18

Adwan, J. (2016). Dynamic online peer evaluations to improve group assignments in nursing e-learning environment. *Nurse Education Today*, *41*, 67–72. doi:10.1016/j.nedt.2016.03.028 PMID:27138485

Adya, M., Temple, B. K., & Hepburn, D. M. (2015). Distant yet Near: Promoting Interdisciplinary Learning in Significantly Diverse Teams through Socially Responsible Projects. *Decision Sciences Journal of Innovative Education*, *13*(2), 121–149. doi:10.1111/dsji.12058

Ali, M. F., Tahir, L. M., & Said, M. N. H. M. (2017). Facilitating learners' online discussions in completing group assignments. In *2016 IEEE Conference on e-Learning, e-Management and e-Services, IC3e 2016*. Institute of Electrical and Electronics Engineers Inc.

Ali, M. F., Tahir, L. M., Said, M. N. H. M., Junaidi, J., Atan, N. A., & Hahsan, A. (2018). Promoting staged self-directed learning (SSDL) among Malaysian tertiary learners through online discussion in completing group assignment. In *2017 IEEE International Conference on Teaching, Assessment and Learning for Engineering, TALE 2017*. Institute of Electrical and Electronics Engineers Inc.

Angelva, J., Tepsa, T., & Mielikäinen, M. (2017). Team teaching experiences in engineering education a project-based learning approach. In *45th Annual Conference of the European Society for Engineering Education, SEFI 2017*. European Society for Engineering Education (SEFI).

Arashpour, M., & Aranda-Mena, G. (2017). Curriculum renewal in architecture, engineering, and construction education: Visualizing building information modeling via augmented reality. In *9th International Structural Engineering and Construction Conference: Resilient Structures and Sustainable Construction, ISEC 2017*. ISEC Press.

Arashpour, M., Bai, Y., Aranda-mena, G., Bab-Hadiashar, A., Hosseini, R., & Kalutara, P. (2017). Optimizing decisions in advanced manufacturing of prefabricated products: Theorizing supply chain configurations in off-site construction. *Automation in Construction*, *84*, 146–153. doi:10.1016/j.autcon.2017.08.032

Arashpour, M., Bai, Y., Kamat, V., Hosseini, R., & Martek, I. (2018). Project production flows in off-site prefabrication: BIM-enabled railway infrastructure. In *35th International Symposium on Automation and Robotics in Construction and International AEC/FM Hackathon: The Future of Building Things, ISARC 2018*. International Association for Automation and Robotics in Construction I.A.A.R.C).

Arashpour, M., Kamat, V., Bai, Y., Wakefield, R., & Abbasi, B. (2018). Optimization modeling of multi-skilled resources in prefabrication: Theorizing cost analysis of process integration in off-site construction. *Automation in Construction, 95*, 1–9. doi:10.1016/j.autcon.2018.07.027

Arashpour, M., Miletic, M., Williams, N., & Fang, Y. (2018). Design for manufacture and assembly in off-site construction: Advanced production of modular façade systems. In *35th International Symposium on Automation and Robotics in Construction and International AEC/FM Hackathon: The Future of Building Things, ISARC 2018*. International Association for Automation and Robotics in Construction I.A.A.R.C).

Arashpour, M., Sagoo, A., Wingrove, D., Maqsood, T., & Wakefield, R. (2015). Single capstone or multiple cornerstones? Distributed model of capstone subjects in construction education. In *Proceedings of 8th International Structural Engineering and Construction Conference, ISEC 2015*. ISEC Press.

Arashpour, M., Too, E., & Le, T. (2017). Improving productivity, workflow management, and resource utilization in precast construction. In *9th International Structural Engineering and Construction Conference: Resilient Structures and Sustainable Construction, ISEC 2017*. ISEC Press.

Arashpour, M., Wakefield, R., Abbasi, B., Arashpour, M., & Hosseini, R. (2018). Optimal process integration architectures in off-site construction: Theorizing the use of multi-skilled resources. *Architectural Engineering and Design Management, 14*(1-2), 46–59. doi:10.1080/17452007.2017.1302406

Asok, D., Abirami, A. M., Angeline Cv, N., & Lavanya, R. (2017). Active learning environment for achieving higher-order thinking skills in engineering education. In *4th IEEE International Conference on MOOCs, Innovation and Technology in Education, MITE 2016*. Institute of Electrical and Electronics Engineers Inc.

Awatramani, M., & Rover, D. (2015). Team-based learning course design and assessment in computer engineering. In *2015 IEEE Frontiers in Education Conference, FIE 2015*. Institute of Electrical and Electronics Engineers Inc.

Beylefeld, A., & Le Roux, A. (2015). Guided group reflections of first-year pre-service teachers: Moving beyond the rhetoric of "go and reflect". *Perspectives in Education, 33*(3), 1–19.

Botterill, M., & Harpe, B. (2010). Working successfully in university interdisciplinary teams: Learning from embedded intergroup relations theory. International Perspectives on Higher Education Research, 5, 75-96.

Brewer, P. E., Mitchell, A., Sanders, R., Wallace, P., & Wood, D. D. (2015). Teaching and learning in cross-disciplinary virtual teams. *IEEE Transactions on Professional Communication, 58*(2), 208–229. doi:10.1109/TPC.2015.2429973

Brooks, N., & Moriarty, A. (2006). Development of a practice learning team in the clinical setting. Nursing Standard, 20(33), 41-44.

Bui, H. T. M., Baruch, Y., Chau, V. S., & He, H. W. (2016). Team learning: The missing construct from a cross-cultural examination of higher education. *Asia Pacific Journal of Management, 33*(1), 29–51. doi:10.100710490-015-9426-z

Caple, H., & Bogle, M. (2013). Making group assessment transparent: What wikis can contribute to collaborative projects. *Assessment & Evaluation in Higher Education, 38*(2), 198–210. doi:10.1080/02602938.2011.618879

Chapman, K. J., Meuter, M., Toy, D., & Wright, L. (2006). Can't we pick our own groups? The influence of group selection method on group dynamics and outcomes. *Journal of Management Education, 30*(4), 557–569. doi:10.1177/1052562905284872

Chen, C. C., Wu, J., & Hayashi, A. M. (2006). Importance of diversified leadership roles in improving team effectiveness in a virtual collaboration learning environment. *12th Americas Conference on Information Systems, AMCIS 2006*.

Clapp, A. (2017). An e-learning team's life on and offline: A collaborative self-ethnography in postgraduate education development. *Electronic Journal of e-Learning, 15*(1), 33-45.

Coccoli, M., Stanganelli, L., & Maresca, P. (2011). Computer Supported Collaborative Learning in software engineering: (ETC - Enforcing Team cooperation). *2011 IEEE Global Engineering Education Conference, EDUCON 2011*. 10.1109/EDUCON.2011.5773267

Crawford, R., & Jenkins, L. (2017). Blended learning and team teaching: Adapting pedagogy in response to the changing digital tertiary environment. *Australasian Journal of Educational Technology, 33*(2), 51–72.

Debuse, J. C. W., & Lawley, M. (2016). Benefits and drawbacks of computer-based assessment and feedback systems: Student and educator perspectives. *British Journal of Educational Technology, 47*(2), 294–301. doi:10.1111/bjet.12232

Denson, N., & Zhang, S. (2010). The impact of student experiences with diversity on developing graduate attributes. *Studies in Higher Education, 35*(5), 529–543. doi:10.1080/03075070903222658

Elahi, B. (2017). Implementation of a learning assessment: Case study of applied statistics course in industrial engineering. In *124th ASEE Annual Conference and Exposition*. American Society for Engineering Education. 10.18260/1-2--27927

Feldmann, B. (2006). Group types in E-learning environments - Study team, working team and learning team. In *7th International Conference on Information Technology Based Higher Education and Training*. ITHET.

Ficano, C. K. C. (2018). Identifying differential benefits from a flipped-group pedagogy in introductory microeconomics. *International Review of Economics Education*.

Francis, K., Henderson, M., Martin, E., Saul, K., & Joshi, S. (2018). Collaborative teaching and interdisciplinary learning in graduate environmental studies. *Journal of Environmental Studies and Sciences*, *8*(3), 343–350. doi:10.100713412-017-0467-0

Fuji, T., Tanigawa, T., Inui, M., & Saegusa, T. (1996). CAMELOT: Collaborative and multimedia environment for learners on teams. *Education and Information Technologies*, *1*(3-4), 203–226. doi:10.1007/BF02350659

Goodman, H., Knight, C., & Khudododov, K. (2014). Graduate Social Work Students' Experiences with Group Work in the Field and the Classroom. *Journal of Teaching in Social Work*, *34*(1), 60–78. doi:10.1080/08841233.2013.866615

Goth, U. S., Bergsli, O., & Johanesen, E. M. (2017). Integration of ethnic minorities during group-work for vocational teachers-in-training in health studies. *International Journal of Medical Education*, *8*, 31–36. doi:10.5116/ijme.5862.505a PMID:28132033

Guilbaud, P., & Whitney, M. (2018). Technology-enhanced learning and the multi-cultural classroom: Exploring impacts of open educational resources' use on collaboration and teamwork. In *6th International Conference on Information and Communication Technology and Accessbility, ICTA 2017*. Institute of Electrical and Electronics Engineers Inc.

Hampton, D. C., & El-Mallakh, P. (2017). Opinions of online nursing students related to working in groups. *The Journal of Nursing Education*, *56*(10), 611–616. doi:10.3928/01484834-20170918-06 PMID:28972631

Han, S. J., Liau-Hing, C., & Beyerlein, M. (2016). Facilitating multicultural student team engagement in higher education: A model for digital learning environments. In Handbook of Research on Transformative Digital Content and Learning Technologies. IGI Global.

Harding, L. M. (2018). Students of a Feather "Flocked" Together: A Group Assignment Method for Reducing Free-Riding and Improving Group and Individual Learning Outcomes. *Journal of Marketing Education*, *40*(2), 117–127. doi:10.1177/0273475317708588

Herr, C. M. (2013). Qualitative structural design education in large cohorts of undergraduate architecture students. *Global Journal of Engineering Education*, *15*(2), 96–102.

Hessler, K. L., & Henderson, A. M. (2013). Interactive learning research: Application of cognitive load theory to nursing education. *International Journal of Nursing Education Scholarship*, *10*(1), 1–9. doi:10.1515/ijnes-2012-0029 PMID:23813334

Howe, T. H., Sheu, C. F., & Hinojosa, J. (2018). Teaching theory in occupational therapy using a cooperative learning: A mixed-methods study. *Journal of Allied Health*, *47*(1), 66–71. PMID:29504022

Hu, H. (2015). Building virtual teams: Experiential learning using emerging technologies. *E-Learning and Digital Media*, *12*(1), 17–33. doi:10.1177/2042753014558373

Huff, P. L. (2014). The Goal Project: A Group Assignment to Encourage Creative Thinking, Leadership Abilities and Communication Skills. *Accounting Education*, *23*(6), 582–594. doi:10.1080/09639284.2014.974198

Hyldegård, J., Hertzum, M., & Hansen, P. (2015). Studying collaborative information seeking: Experiences with three methods. In Collaborative Information Seeking: Best Practices, New Domains and New Thoughts. Springer International Publishing.

Joshua, E., & Mariajose, C. (2013). *A comparison of peer evaluation methods in capstone design. 120th ASEE Annual Conference and Exposition*, Atlanta, GA.

Kalfa, S., & Taksa, L. (2017). Employability, managerialism, and performativity in higher education: A relational perspective. *Higher Education*, *74*(4), 687–699. doi:10.100710734-016-0072-2

Kampe, J., & Oppliger, D. E. (2011). Effectiveness of team-based STEM project learning to recruit minority high school students to STEM. In *ASEE Annual Conference and Exposition, Conference Proceedings*. American Society for Engineering Education.

Keevers, L., Lefoe, G., Leask, B., Sultan, F. K. P. D., Ganesharatnam, S., Loh, V., & Lim, J. S. Y. (2014). 'I like the people I work with. Maybe I'll get to meet them in person one day': Teaching and learning practice development with transnational teaching teams. *Journal of Education for Teaching*, *40*(3), 232–250. doi:10.1080/02607476.2014.903024

Khosa, D. K., & Volet, S. E. (2013). Promoting effective collaborative case-based learning at university: A metacognitive intervention. *Studies in Higher Education, 38*(6), 870–889. doi:10.1080/03075079.2011.604409

Khuzwayo, M. E. (2018). Assessment of group work in initial teacher education and training. *South African Journal of Education, 38*(2).

Lalopa, J. M., Jacobs, J. W. Jr, & Hu, C. (1999). Evolving learning environments: Another team-based example in higher education. *Journal of Hospitality & Tourism Education, 10*(4), 30–37. doi:10.1080/10963758.1999.10685207

Lambert, S. C., Carter, A. J., & Lightbody, M. (2014). Taking the guesswork out of assessing individual contributions to group work assignments. *Issues in Accounting Education, 29*(1), 169–180. doi:10.2308/iace-50637

Lee, H. J., Kim, H., & Byun, H. (2017). Are high achievers successful in collaborative learning? An explorative study of college students' learning approaches in team project-based learning. *Innovations in Education and Teaching International, 54*(5), 418–427. doi:10.1080/14703297.2015.1105754

Magogwe, J. M., Ntereke, B., & Phetlhe, K. R. (2015). Facebook and classroom group work: A trial study involving University of Botswana Advanced Oral Presentation students. *British Journal of Educational Technology, 46*(6), 1312–1323. doi:10.1111/bjet.12204

Maina, E. M., Oboko, R. O., & Waiganjo, P. W. (2017). *Extending moodle grouping functionality using artificial intelligent techniques. In IEEE AFRICON 2017.* Institute of Electrical and Electronics Engineers Inc.

Mangum, S. W., Johnson, J. L., Gunaldo, T. P., Casanova, E., & DeKerlegand, A. (2017). Developing student interprofessional competence through a support group assignment. *Journal of Interprofessional Education and Practice, 9*, 17–20. doi:10.1016/j.xjep.2017.07.006

Maqbool, J. (2016). Group work and the impact, if any, of the use of google applications for education. *International Conference on Cognition and Exploratory Learning in Digital Age.*

Matveev, A. V., & Milter, R. G. (2010). An implementation of active learning: Assessing the effectiveness of the team infomercial assignment. *Innovations in Education and Teaching International, 47*(2), 201–213. doi:10.1080/14703291003718935

Mendo-Lázaro, S., León-del-Barco, B., Felipe-Castaño, E., Polo-del-Río, M. I., & Iglesias-Gallego, D. (2018). Cooperative team learning and the development of social skills in higher education: The variables involved. *Frontiers in Psychology*, *9*. PMID:30186208

Meschitti, V. (2018). Can peer learning support doctoral education? Evidence from an ethnography of a research team. *Studies in Higher Education*, 1–13.

Moore, P., & Hampton, G. (2015). 'It's a bit of a generalisation, but …': Participant perspectives on intercultural group assessment in higher education. *Assessment & Evaluation in Higher Education*, *40*(3), 390–406. doi:10.1080/02602938.2014.919437

Nasirian, A., Arashpour, M., & Abbasi, B. (2018). Multiskilled human resource problem in off-site construction. In *35th International Symposium on Automation and Robotics in Construction and International AEC/FM Hackathon: The Future of Building Things, ISARC 2018*. International Association for Automation and Robotics in Construction I.A.A.R.C).

Ng, W., & Hanewald, R. (2009). Concept maps as a tool for promoting online collaborative learning in virtual teams with pre-service teachers. In Handbook of Research on Collaborative Learning Using Concept Mapping. IGI Global.

Noguera, I., Guerrero-Roldán, A. E., & Masó, R. (2018). Collaborative agile learning in online environments: Strategies for improving team regulation and project management. *Computers & Education*, *116*, 110–129. doi:10.1016/j.compedu.2017.09.008

Odeh, S., McKenna, S., & Abu-Mulaweh, H. (2017). A unified first-year engineering design-based learning course. *International Journal of Mechanical Engineering Education*, *45*(1), 47–58. doi:10.1177/0306419016674508

Poelmans, S., & Wessa, P. (2015). A constructivist approach in a blended e-learning environment for statistics. *Interactive Learning Environments*, *23*(3), 385–401. doi:10.1080/10494820.2013.766890

Rae, D., Gee, S., & Moon, R. (2009). Creating an Enterprise Culture in a University: The Role of an Entrepreneurial Learning Team. *Industry and Higher Education*, *23*(3), 183–197. doi:10.5367/000000009788640279

Scager, K., Boonstra, J., Peeters, T., Vulperhorst, J., & Wiegant, F. (2016). Collaborative learning in higher education: Evoking positive interdependence. *CBE Life Sciences Education*, *15*(4), ar69. doi:10.1187/cbe.16-07-0219 PMID:27909019

Schaber, P., McGee, C., & Jones, T. (2015). Building student community in a hybrid program. *Occupational Therapy in Health Care*, *29*(2), 102–114. doi:10.3109/07380577.2015.1017134 PMID:25821882

Scott, J. L., Mistry, S., Moxham, B. J., & Rutherford, S. M. (2014). Using web 2.0 technology to support and enhance collaborative activity outside of the taught curriculum in higher education. In Collaborative Learning: Theory, Strategies and Educational Benefits. Nova Science Publishers, Inc.

Serçe, F. C., Nur Alpaslan, F., Swigger, K., Brazile, R., Dafoulas, G., & Lopez, V. (2010). Strategies and guidelines for building effective distributed learning teams in higher education. *2010 9th International Conference on Information Technology Based Higher Education and Training*.

Skelley, J. W. (2014). A small group assignment gives students a novel opportunity to demonstrate current clinical controversies in a self-care course. *American Journal of Pharmaceutical Education*, *78*(10), 1–8. doi:10.5688/ajpe7810193 PMID:25657380

Smith, M. A., & Kellogg, D. L. (2015). Required Collaborative Work in Online Courses: A Predictive Modeling Approach. *Decision Sciences Journal of Innovative Education*, *13*(4), 563–581. doi:10.1111/dsji.12078

Stache, J. M., Barry, B. E., Klosky, J. L., & Freisinger, G. M. (2017). Assessing individual temperament and group performance in a project-based learning experience. In *124th ASEE Annual Conference and Exposition*. American Society for Engineering Education. 10.18260/1-2--27621

Streiner, S., Besterfield-Sacre, M., & Donovan, S. (2016). Using simulation and structured group work to address statistical misconceptions. In *46th Annual Frontiers in Education Conference, FIE 2016*. Institute of Electrical and Electronics Engineers Inc.

Thrall, G. C., Coverdale, J. H., Benjamin, S., Wiggins, A., Lane, C. J., & Pato, M. T. (2016). A Randomized Controlled Trial of Team-Based Learning Versus Lectures with Break-Out Groups on Knowledge Retention. *Academic Psychiatry*, *40*(5), 755–760. doi:10.100740596-016-0501-7 PMID:26883529

Van Leeuwen, A. (2016). Learning analytics in a flipped University course. In *6th International Conference on Learning Analytics and Knowledge, LAK 2016*. Association for Computing Machinery. 10.1145/2883851.2883874

Wang, E. L., & Vollstedt, A. M. (2014). A method for adjusting group-based grades. In *121st ASEE Annual Conference and Exposition: 360 Degrees of Engineering Education*. Indianapolis, IN: American Society for Engineering Education.

Wang, J., Tsiligiris, V., & Hartley, R. (2018). Perceptions of the benefits of the A3 planner in facilitating project-based learning in accounting education. *Accounting Education.*

Woolard, N. A. (2018). Rethinking management group projects with the adaptive leadership model: The lesson is the process. *Journal of Education for Business, 93*(8), 392–402. doi:10.1080/08832323.2018.1496895

Yannibelli, V., & Amandi, A. (2017). A hybrid evolutionary algorithm based on adaptive mutation and crossover for collaborative learning team formation in higher education. *18th International Conference on Intelligent Data Engineering and Automated Learning.*

Yuretich, R. F., & Kanner, L. C. (2015). Examining the effectiveness of team-based learning (TBL) in different classroom settings. *Journal of Geoscience Education, 63*(2), 147–156. doi:10.5408/13-109.1

Zitzelsberger, H., Campbell, K. A., Service, D., & Sanchez, O. (2015). Using wikis to stimulate collaborative learning in two online health sciences courses. *The Journal of Nursing Education, 54*(6), 352–355. doi:10.3928/01484834-20150515-08 PMID:26057431

Zschocke, K., Wosnitza, M., & Bürger, K. (2016). Emotions in group work: Insights from an appraisal-oriented perspective. *European Journal of Psychology of Education, 31*(3), 359–384. doi:10.100710212-015-0278-1

Chapter 4
Learning Through Successful Digital Opportunities for Effective Competition Preparations

Lloyd Martin Scott
https://orcid.org/0000-0001-6305-3749
Technological University Dublin, Ireland

Mark Shelbourn
University of Salford, UK

Nicky Harris
University of Salford, UK

ABSTRACT

Education of the built environment is moving towards more collaborative practices. The intent behind the collaborative approach of teaching is to encourage students to explore the unknowns and unravel the problems themselves with the professor acting as the facilitator. This chapter presents the case for collaborative pedagogical approaches that was adopted to teach students from two geographically distant universities. The occasion used was preparation of student teams for Associated Schools of Construction (ASC) student competitions. The underlying philosophy behind this initiative is to provide the context for the future leaders in the AEC to learn by doing in a safer and potentially rewarding environment.

DOI: 10.4018/978-1-5225-8452-0.ch004

INTRODUCTION

There is a growing demand for closer collaboration within the design and construction disciplines in the Built Environment. Penttila and Elger (2006), suggest that diverse multidisciplinary understanding and knowledge about various factors of design and construction will be essential in the near future (architectural) design profile and so to provide for industry. There has always been a collaboration of sorts in the Architecture, Engineering and Construction (AEC) sector, this has been more based on co-operation than true collaboration in most cases (Tran, Nguyen &Faught, 2017) Buildings would not get built without the inputs from the various stakeholders but it could be defined more as co-operation based on the requirement of necessity. The co-operation has tended to be cautious and protective of one's information and much has been based on the culture with the AEC sector. This co-operation as opposed to collaboration has led to miscommunication within the process. The apportioning of the 'blame culture' has resulted in a litigious, conflict driven and difficult process. BIM promotes methodologies of integrated design, integrated project delivery and signposts towards closer collaboration as opposed to co-operation. There are case studies which show a major improvement in building delivery where better communication of design and construction information have led to much less problems on site. The case study will be referred to later in this research study.

If this is the case and BIM technologies and methodologies can and are providing an alternate to this traditional process in the AEC industry then it is legitimate that these technologies and methodologies should to be reflected back into the teaching and learning practices of educators in the disciplines of design, measurement and construction. How will this benefit both the educators and the students? What effects will these new process have on the pedagogy of teaching and learning in Higher Education Institutions. One of the challenges higher education institutions will face is to break down silos of built environment education and provide opportunities for the development of collaborative skills for students poised to enter into the design and construction industry.

Will educators need to create new learning environments to fully utilise the potential of BIM as a learning tool? What will be the benefits of these new learning environments for both educators and students? How will these new learning environments be structured so as to achieve the learning outcomes for the new roles in the industry?

Building upon previous research which set out to observe and record a collaborative design project taught in the Department of Construction Management at the Dublin Institute of Technology (DIT) the author observed a developing learning environment that accelerated the students learning. Literature provides evidence of effective collaborative practise in the industry, further strengthening the case for both BIM and team member based collaborative learning in higher education. What has begun to come to the surface is greater investment by academics in providing opportunities for more authentic learning opportunities (Scott, 2012).

BACKGROUND

McGraw-Hill have published a number of reports where they asked North American Architecture Engineering & Construction (AEC) firms their requirements for skills for collaborative BIM. In 2009 (McGraw Hill, 2009), they reported that "...more internal staff with BIM skills, more external firms with BIM skills, more incoming entry-level staff with BIM skills and more readily available training in BIM were required to realize the potential value of BIM." By 2012 (McGraw Hill, 2012) the updated report showed small decreases in the percentages allocated to the collaborative BIM skills required (possibly reflecting uptake by the industry), but collaborative BIM training was still placed among the top three targets for investment by industry.

These reports show similarities with Henderson and Jordan (2009) who suggested that some of the additional skill-sets (to their traditional single-discipline learning) industry requires included: "...knowledge of data management, information technology, energy and material conservation, integrated building design, systems thinking, life cycle analysis, the design processes, business and marketing skills, and project finance" (p. 35).

It is the role of educators to instil in all graduates the concepts of collaborative design and the full potential of collaborative team integration, before they learn about the "*old ways*" of working once they graduate (Shelbourn et al. 2017). The concept of creating job-ready graduates brings to the fore the "training vs. educating" debate. Gerber *et al.,* (2013) demonstrated that there has been resistance in the past among educators in providing training in collaborative computer technologies in universities, as many educators are unfamiliar with such technologies. This often means educators currently expect students to learn it by themselves, as they do many other software applications (Williams et al., 2009). Given these precedents, one can assume the same approach to learning collaborative BIM meaning students will tend to focus on the technological aspects, rather than develop an understanding of how BIM principles and processes could enable them to work more effectively with others in a collaborative team environment.

As far as the training v's education debate is concerned, many educators still view BIM as just another piece of Computer-Aided-Drafting (CAD) software that students should learn in their own time. Many argue that it is not the university's role to produce 'CAD technicians' and that there is little educational value in using CAD, or that CAD threatens creativity (Becerik-Gerber *et al.*, 2011). These concerns can be justified as the adoption of computers and 2D CAD has coincided with a decrease in documentation quality and productivity (Engineers Australia, 2005).

However, collaborative BIM is not merely a new CAD tool or computer application: it is a new paradigm and its benefits extend much further than 3D drafting (Chegu Badrinath *et al.* 2016). Students cannot be expected to "teach themselves BIM" any more than they could be expected to "teach themselves structural engineering" (Engineers Australia, 2005; Gledson *et al.* 2016). From a learning point of view, there is little difference between learning manual drafting techniques and learning 2D or 3D CAD. However, with collaborative BIM, every part of the design and construction process and can be compared, with building performance also modelled at this stage and monitored in the operation phase. Both 2 and 3D CAD merely provide a way of documenting information about the building, whereas collaborative BIM actually represents the building virtually with critical information contained within it to help optimise the operation of the facility throughout its lifecycle (Hu *et al.* 2017).

In addition to the resistance to using new technologies in teaching, faculties where this learning is taking place is also a barrier to learning (Shelbourn *et al.* 2016). Since engineering and architecture emerged as separate professions from the historical job title of Master Builder, students of the different disciplines have tended to be educated in isolation from each other. According to Pressman (2007, p., 3):

Many academic programs still produce students who expect they will spend their careers working as heroic, solitary designers. But integrated practice is sure to stimulate a rethinking of that notion. Pedagogy must focus on teaching not only how to design and detail, but also how to engage with and lead others, and how to collaborate with the professionals they are likely to work with later.

Starzyk and McDonald (2010) reported a focus of architectural education on developing individual skills, such as the ability to draw. More recently, they state, "… the importance of personal skill is yielding to the primacy of collective knowledge". Scott (2015) found little or no integration or collaboration between the disciplines in the majority of universities in the USA, Europe and Australia. Moreover, the first time that students are exposed to working with team members from other disciplines is in the workplace, post-graduation. Shelbourn *et al.* (2017) discussed this further and said "…it is important for graduates to have an understanding of the roles played by other professionals and the impact their decisions have on projects overall." However, the lack of multi-disciplinary collaborative learning means students are not provided with such an understanding in much of the current curricula across these countries.

Another issue to consider is the complexity of modern building projects and technologies used in their design and construction. Such complexity means that nobody can be a master of all. Students learning in their silos lack a deep understanding of the information that each requires at different stages of a project (Shelbourn *et al.* 2017). What is required is for students to work collaboratively and to learn the requirements of the other disciplines before they graduate, often in multi-disciplinary modules, projects and even student competitions such as those offered as part of the Associated Schools of Construction in the USA.

The problem is not only restricted to students of separate disciplines learning in their own silos: different departments are often in separate departments, schools or faculties and can be located on separate campuses (Shelbourn *et al.* 2016). Sharing learning across the different silos is a challenge that needs addressing if graduates are to leave their studies with the key skills of understanding the importance of collaboration (Shelbourn *et al.*, 2016). The need for change instigated by the BIM revolution (Cabinet Office, 2011) provides a great opportunity to rethink how teaching and learning is designed with educators becoming more efficient in delivering it (Shelbourn *et al.* 2017).

It is not all doom and gloom however. Hardy, quoted in Deutsch (2011, p. 202) stated, "...when I look at the logic of construction means and methods that collaborative BIM inherently teaches, I see the potential to educate...". Nawari (2010) noted that "...students need to know how each discipline is related to the other and how one discipline impacts the other". The role of collaborative BIM can offer a better opportunity to engage students more effectively and to help with their understanding of how buildings are constructed.

Mark *et al.* (2001) proposed "...the ideal computer curriculum..." for architectural education where computing technologies were added to existing curriculum without removing or adding subjects. Mark *et al.* (2001) offered two alternatives: one that merged technology into the traditional curriculum, the other a more radical approach that displaced some existing subjects. The approach was limited to teaching BIM modelling for visualization or analysis within the architectural discipline alone. Scott (2016, p.552) highlighted the case for setting education in the pragmatic paradigm, pointing out that "...the freedom to work within the pragmatic paradigm offers diversity that can draw together some of the thoughts that challenge and build the arguments about the role and position of theory in construction education" – certainly a useful consideration when looking at collaborative BIM education.

The challenge for academics wanting to educate undergraduates so that they can work effectively in collaborative teams, putting together virtual (and eventually real-life) buildings, is *when* and *how* to introduce elements of disciplinary knowledge, BIM technologies and development of team working skills. Collaborative BIM education should be effected in stages (Shelbourn *et al.* 2016), increasing in complexity as the students' knowledge of the building design and construction process grows (Gordon *et al.*, 2009).

According to Vygotsky (1978), students are capable of performing at higher intellectual levels when asked to work in collaborative situations. This was one of the primary encouragements for completing this project. It is thought within both the Construction Management program and the Quantity Surveying program that collaboration projects will have a positive effect on the students learning. It should also be notes that on graduation the students will be required to work alongside many other professions. We felt if we could create a project that brings students together to work in an interactive manner; it may in some way reflect their work environment into the future. The current economic climate has diminished employment prospects for students from both programs in Ireland, but the potential for employment exists outside of Ireland in particular for those who have enhanced skills in new technology and communications (ICT). New technology and communication skills will greatly improve employment prospects of students from both these programs

The long term strategic aim of this project was to expand the opportunity into other programmes and to introduce a collaborative project that would allow all the key professions within a construction project to contribute to the project. It would be the intention to bring in Architecture, Engineering and property students in future collaborative projects. It is proposed that this project will act as a model for future projects. It is envisaged that the two dimensional drawings and the Building Information Model will be used again in future years for the collaborative projects. Once the software demands are met in the first year of this project, the demand on resources will be greatly reduced in future years.

Solving a Distinct Problem in Multidisciplinary Built Environment Education

The action research that is presented in this chapter involved cycles of interactions between the authors and participants focusing on two separate objectives:

1. To enhance participants' learning to build capabilities and abilities to understand and generate constructive behavioral change through reflective action on assessment practice
2. To evaluate the usefulness of collaborative learning to enhance such actions by the participants as improving their learning.

Action Research

Action research was deemed as the ideal methodology when a holistic, in-depth investigation was the goal (Feagin, Orum & Sjoberg 1991). While action research is distinguished by its purpose that is to affect change in the participants, the objective of the ASC student competition (specifically the 'International Design Build' competition) was to enhance the participants' capabilities (Scott, 2016). The study presented in this chapter did not attempt to perform sampling research. Within the premise of the approach adopted in this research, there is a choice to be made in relation to the number of cases to include and as the initiative will be ongoing it will very much be considered as presenting a number of case studies over time, the occurrence in consideration being the first.

The growing use of action research within built environment education research and development initiatives explicitly recognize that collaborative approaches to student learning are not characterized so much by problems for which an answer must be found, but rather by issues which need to be resolved and will inevitably require one or more of the parties to change their views. The underlying assumption of these approaches is that effective social change depends on the commitment and understanding of those involved in the change process. In other words, if people work together on a common problem 'clarifying and negotiating' ideas and concerns, they will be more likely to change their minds if their 'joint research' indicates such change is necessary. Also, it is suggested that collaboration can provide people with the interactions and support necessary to make fundamental changes in their practice which endure beyond the research process. Action research comprises a family of research methodologies, which aim to pursue action and research outcomes at the same time. It therefore has some components which resemble consultancy or change agency, and some which resemble field research. The focus is action to improve a situation and the research is the conscious effort, as part of the process, to formulate public knowledge that adds to theories of action that promote or inhibit learning in behavioral systems. One of the key characteristics of this approach is collaboration, which enables mutual understanding and consensus, democratic decision making, and common action (Oja & Smulyan, 1989).

In this sense the action researcher is a practitioner, an interventionist seeking to help improve client systems. "This help takes the form of creating conditions in the behavioral world of the client system that are conducive to inquiry and learning. Lasting improvement requires that the participatory action researcher help clients to change themselves so that their interactions will create these conditions for inquiry and learning" (Argyris et al., 1985 p.137). Hence to the aims of contributing to the practical improvement of situations and to the goals of developing knowledge we can add a third aim of action research, to develop the self-help competencies of people facing problems. Within this broad definition there are four basic themes: (1) collaboration through participation, (2) acquisition of knowledge, (3) social change, and (4) empowerment of participants. The process that the researcher uses to guide those involved can be seen as a spiral of action research cycles consisting of phases of planning, acting, observing and reflecting (Masters 1995) and identified in Figure 2. As Oja and Smulyan (1989) point out, the underlying assumption of this approach, which can be traced back to Lewin's (1948) writing that effective social change depends on the commitment and understanding of those involved in the change process. In other words, if people work together on a common problem 'clarifying and negotiating ideas and concerns,' they will be more likely to change their minds if research indicates such change is necessary. Also, it is suggested that collaboration can provide people with the time and support necessary to make fundamental changes in their practice which extends beyond the research process (Oja & Smulyan, 1989). Thus, the role of the action researcher is identical to that proposed for contemporary facilitators in helping communities identify and adopt more sustainable natural resource management practices (Pyecha, 1988). The adoption of the Kemmis and McTaggart (2000) action research approach (see Figure 1) is a participatory study consisting of spiral of following self-reflective cycles:

1. Planning in order to invoke change
2. Implementing that change and then observing the process of implementation and the consequences of that change
3. Reflecting on the process of change and then re-planning
4. Acting and observing
5. Reflecting.

Figure 1. The Kemmis and McTaggart (2000) spiral action research process

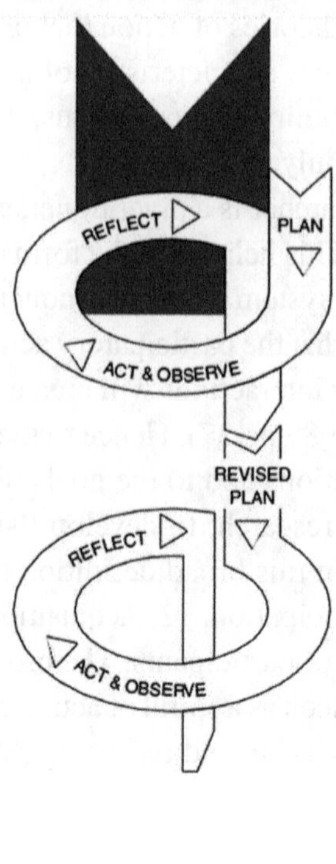

IMPORTANCE OF COLLABORATION AS A GRADUATE ATTRIBUTE

The Importance of Collaborative Learning

Whether an educator facing a lecture hall filled with 300 students or a seminar table with 15 students, one of the primary goals for the event should be to actively engage students with the required learning material. Students learn more when they participate in the process of learning, whether it is through discussion, practice, review, or application (Grunert O'Brien, Millis, Cohen, 2009). This is in stark contrast to traditional styles of teaching, where students are expected to sit for hours, listening and, theoretically, absorbing information presented by the instructor.

Collaborative Education in the Built Environment

Incorporate active and collaborative learning strategies into every component of your course/ programme or project design. For example, encouraging short partner discussions during preparation sessions (i.e., think-pair-share), adding problem-or case-based research projects to the project as preparation, and incorporating time for small-group critical analysis exercises during team seminars are all great ways to actively engage students in learning.

Because it can take time and creativity to develop these type of learning tasks/ exercises, cognizance should be taken to plan in the necessary timing.

Facilitate Independent, Critical, and Creative Thinking

Ask students to analyze, synthesize, or apply material, both during preparation sessions and in individual assignments. Some examples include:

- **Case-Based Problem Solving Exercises**: These types of exercises helped students develop analytical skills and learn how to apply academic theories to real-world problems. Use case studies in an early practice stages and have students work out their solutions independently and then in the small group was very positive. Doing them remotely did provide some challenges especially when it cam to do a simulated practice.
- **Debate/ Discussion:** This proved to be a really positive active learning technique that helped develop and improve critical thinking and logical reasoning skills within the team. The offering of competing viewpoints and encouraging students to defend one, or both, of the viewpoints in a short

debate like scenarios proved successful as the team(s) built thrust and belief as part of that engagement.

Encourage Effective Collaboration

Collaborative group work by way of asking the team members to prepared a presentation to the 'client' was a method used in the practice time before the competition. Research from cognitive psychology has shown that one of the best ways to improve understanding is to teach/explain material to a peer (Topping & Stewart, 1998). Build this exercise into the learning experience of the student was achieved through presentations, study groups, and quick, breakout "teaching" sessions as was the approach adopted for the "Design Build" experience.

Increase Student Investment, Motivation, and Performance

When you invite students to actively and collaboratively participate in the learning environment, they take more responsibility for their performance in the course/ task or project. Similarly, when they have an opportunity to make decisions about what they learn and how they use that knowledge, students see a task/project as more valuable and more directly related to their goals. For example, the time spent at the start of a project can be eroded away easily if a team did not prepare effectively. A method used to encourage quality time planning is building in time to map out the objectives of the project.

- **Brainstorm Learning Objectives:** If you involve students in the development of team activities, e.g., allow them to choose the topic of a short discussion or generate ideas about how a concept could be applied to a problem, it automatically increases engagement levels. Involving team members in the planning activities also requires them to assess their understanding and skill and rather than allowing them to rest comfortably with a surface knowledge, it forces them to develop a deeper understanding of the material/ requirements.

FINDINGS/FEEDBACK FROM PARTICIPANTS

Following on from the planning/ structuring around the PAR approach he authors have collected initial data from 27 participants of the recently completed ASC student competition (specifically the 'International Design Build' competition) to determine the usefulness of the collaborative pedagogical approach. The authors asked several questions to the participants in the form of an online survey. The summary of the responses is presented below. The participants were asked to rate the different aspects towards the preparation of the competition in a three-point scale (very helpful, neutral, and least helpful). Almost all of the participants (94%) mentioned the support from the coaches was most helpful followed by support from the industry guest speakers (Figure 1). A major portion of the learning of the 'International Design Build' team member's entails getting accustomed to the construction and contractual processes of the US construction industry along with their jargons. The authors along with other coaches invited speakers from the industry to bridge this gap. The participants' low agreement on the helpfulness of the other faculty members may be due to preoccupation of the faculty members with other commitments. The virtual team meetings play an important role in the team building process spanning the educational and cultural diversity, and laying the foundation for collaboration. This team cohesiveness and collaboration among the members generate the sense of 'commonness' and the urge to work together for 'clarifying and negotiating' ideas. It has been the experience of the authors that during the week before the actual competition when the team members meet face-to-face, a lot of unlearning and relearning take place. By this time the social change can be seen taking grip on the participants in the form of changing their mind or giving away preconceived notions if their joint research demands so. One of the participants commented:

The virtual team meetings are important but there is a significant lack of the subtle communication that makes meetings important. You can handle generic information and technicalities over skype but it is much easier for misunderstandings or potential issues regarding "closed" matters to go unnoticed. (Team member 23)

Figure 2. Participants agreement with usefulness of different aspects of preparation

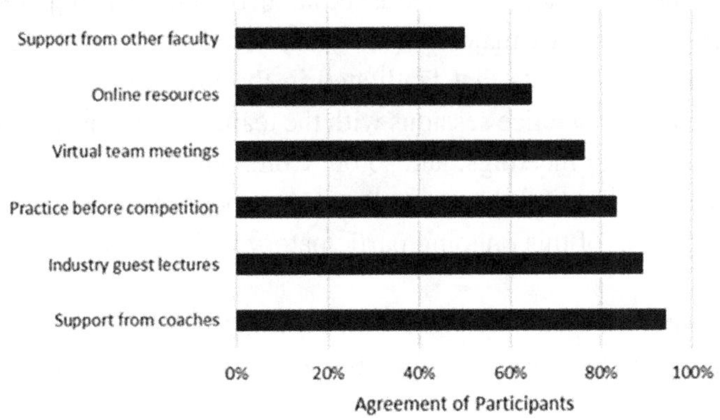

Analysis

The feedback from the participants has identified that the learning context is a significant aspect to learning. Teamwork emerged as a real bonus and to understand the perspective of the other stakeholders was a very meaningful outcome; the architect learning from the construction manager and visa versa.

The personal learning experiences speak for themselves and just a few of the personal reflections are shared here:

From the entire international design/build experience I learned that teamwork comes in all forms. Our team did not meet face to face until 2 days before the competition but we broke the ice very quickly and bonded. For me, the entire experience proved that drive and determination will overcome any task, individually or as a group. (Team member 5)

This project was an inspirational experience, full of fascinating people who exchanged ideas and learned ideas and learned from one another in a rich collaborative process. (Team member 11)

The entire competition and preparations were hard work but Educationally rewarding plus a once in a life time experience. From making friends half way round the world and here at home, getting to travel, getting a glimpse of how the American's do construction, and the really enjoyable time we all had working together as a team. (Team member 17)

The authors in conjunction with other coaches agree that the ASC competitions demand tremendous amount of effort and superior level of maturity from the participants. Preparing for these type of competitions is always challenging and involves some level of unpredictability. Subsequent to the completion of the competition, when the participants were asked about their level of preparedness going into the competition, a vast majority (76%) answered in affirmative. The factors identified by the participants that facilitated in the preparation process included (but not limited to) the practice sessions with the team, the feedback of the coaches, the online and physical meetings, and so on. Comment by one of the participants highlighted the social value, change and empowerment of the participants, which remains the objective of this ongoing participatory action research.

.... each member had specific strengths and experience in their part of the project. These were identified early in the process so preparation could be maximized. (Team member 4)

One of the objectives of the PAR was to evaluate the usefulness of this collaborative pedagogical approach. The authors based their conclusions on two measures: the performance of the student teams in the ASC competitions throughout the event and the individual perceptions of the participants. While the strong performance of the teams provide testimony in favour of the collaborative approach, students engaged with each other and once the early formalities were encountered the cohesion of the team members was very evident.

The indirect measure based on the participants' perceptions were also equally positive. A few comments shown below will demonstrate the positive change in the participants' attitude, knowledge, and competencies.

The teamwork was definitively the best aspect. As part of the international team, we met new people of different nationalities and this was extremely interesting. (Team member 19)

...competitive and very challenging competition that strengthens students for a similar real life situation in the student future career. (Team member 9)

The team worked well together and while this can be expected in a situation there was occasion where tension appeared to be high. The hour leading up to submission time did provide some aspects for reflection and all team members indicated that were they to do the project again then their approach to final submission requirements would be 'handled differently'. More planned out and time set aside for the final preparations.

Indeed, there are many positives to be highlighted for academics and students in pursuing such an innovative approach to learning and teaching. For students, the structured and applied nature of the work they are required to undertake in their virtual groups goes some way towards providing an alternative approach to engaging them in learning through a flexible, enhanced teaching method, on an open-access online platform. It is, for students, a rewarding experience that fits with the current trends in curricular development, which seek to encourage pedagogical models that are innovative, are student-centred, include the use of technology where apposite, are flexible in their delivery (on and off campus), enhance the student experience, encourage them to be more autonomous in their learning, and competent problem solvers in order to raise their aspirations and develop their employability skills.

Lessons Learnt

In terms of offering colleagues in the CM discipline some words of wisdom from the authors reflections, the first comment would be there is no manual for finding our way through this type of initiative. Despite the previous experience in this collaborative approach to enhancing the student experience both academics are very much of the view that this is the students learning experience and so imposing a pre-prepared structure on the team was not going to work. With the practical aspects and personal interactions that this kind of team experience requires a loose structure for the teams to consider and reflect upon is proposed. As might be expected, the process will be quite alien to the groups at the start, so it is important to provide as many resources as possible to the team members.

Some of the fundamental points to consider would be:

- Allow the members of the group to share their knowledge and gain confidence - allow and schedule time for this as this will require more time than you might expect
- Encourage them to schedule meetings outside of the designated times
- Do not rely on the online interaction to be the definitive way to communicate
- Encourage the team members to challenge assumptions
- Ensure that team members, especially those who will have to travel to the competition acquire as much background knowledge as possible
- Embrace the student-led collaborative efforts that lead to team success and look for ways to foster it
- Make any expectations clear to the them members
- Be comfortable with the fact that this is the students experience and so they may encounter some obstacles.

This kind of truly collaborative pedagogical approach demands a major time commitment. One cannot assume that the team members know what it is they are going to say and roll with it as easily. Be prepared to have situations that will take more time than you might have scheduled for. The time spent will allowed team members to deepen their understanding of the requirements to be successful, improve interactions with each other, develop a capacity to embrace differences, and work toward a more collaborative approach to solving the project. This interesting account of two coaches (teachers) who truly collaborated as they jointly taught shows how much teachers can learn when they work together. This endeavor will be time-consuming, but remember it has the potential to provide a commensurate amount of personal growth and development for all concerned. It is certainly recommended to early career academics and is a really solid way of developing high quality learning and teaching pedagogy.

While English is the common language used among team members, there will be some culturally subtle differences. For example, construction terminology can be quite different across the globe and experience has shown the authors that it is only when the students meet face to face that they really get to understand those differences. This holds true for the approach to measurement systems too, non U.S. students tend to need to up skill themselves on the imperial system as they are normally conversant with the metric system only. This often provides for a huge learning curve on the part of the architect and the estimator/ quantity surveyor in particular.

DISCUSSION AND FUTURE DIRECTIONS

The authors reflected on a number of advantages such a collaborative learning format - social benefits, learning benefits, and development of skills, knowledge and competences for the students' future careers. The early stage virtual meeting was embedded to reduce the social anxiety of students by providing an instant group of peers with whom they would not feel exposed. Instead they would feel a sense of community through engaging in the common task of grappling with and understanding the competition structure and the material associated with it. Secondly, it was hoped that the method would help to promote deeper understanding, especially for the international students as their knowledge and experience in the US construction processes was very limited. The upskilling and deep learning around US practice of estimating and health & safety regulations challenged the international students but what did emerge was the experience of sharing of best practice from both processes. While the international students certainly had a steeper learning curve the overall emerging theme was one of a deeply constructive learning experience and one that they participants have indicated would be lasting.

The future directions to consider by both the authors and the construction education community is how such really meaningful learning experiences get embedded in the educational experiences of the learners we come in. Some positive starting points include:

- Continued collaborative engagement
- Share what has been developed to date
- Develop a rotational model that allows students to participate abroad
- Encourage other Universities to participate
- Create shared mission of active learning approach
- Create significant learning opportunities for Built Environment students

Students were exposed to different styles of thinking, collaborating and different ways of tackling the problems. Having already thought about the problems, the arguments and comments of the coaches were more meaningful to the students. Thirdly, the authors thought that interacting in small groups would give students practice in communication and interpersonal skills useful in their later careers. Construction professionals often operate in formal or informal small groups, asking questions and advising peers, sometimes accepting and at other times rejecting ideas. Construction professionals also form their own opinions by synthesizing the contributions of others with their own previous experience and knowledge. The authors thought that the competition format would provide valuable practice in these skills.

The research reported on in this chapter has begun to make a positive impact where this type of pedagogical initiative has been adopted and reflected upon. One suggestion to the academy in the built environment is to start at school/college and try to embed the approach there where Architects, Engineers, Quantity Surveyors and Construction Managers can come together to learn collaboratively. This can only be done with strong education orientated academy focused on authentic learning. Changing the model of pedagogy and the model of knowledge production is crucial at program level for the survival of the university and the programs they offer.

The holistic, interdisciplinary approach to the design, construction, production and operation of buildings is likely to require changes in the way the process is arranged, resourced and managed in the future (Atkin, 2009). There will be a different kind of professional in the twenty or thirty years whose education and/or training will be required to permit them to make the many connections in thinking and actions required to solve complex problems. Future construction professionals must challenge the conventional ways of the past in ways that use their creative and innovative capacities. Active learning experiences like the Design Build competition offer so many opportunities to develop this.

Learning is the process whereby knowledge is created through transformation of experience (Kolb, 1994)

ACKNOWLEDGMENT

The authors would like to acknowledge the participation of the participating students who gave of their time to offer reflections and feedback.

REFERENCES

Ahmed, A. K. (2013). Teacher centered versus learner centered teaching style. *The Journal of Global Business Management.*, *9*(1), 22–34.

Argyris, C. (1985). *Action Science, Concepts, methods, and skills for research and intervention*. San Francisco: Jossey-Bass.

Becerik-Gerber, B., Gerber, D. J., & Ku, K. (2011). The pace of technological innovation in architecture, engineering, and construction education: Integrating recent trends into the curricula. *Journal of Information Technology in Construction*, *16*, 411–432.

Bryman, A., & Bell, E. (2011). *Business Research Methods* (3rd ed.). Oxford University Press.

Cabinet Office. (2011). *Government Construction Strategy*. Available from: https://www.gov.uk/government/uploads/system/uploads/attachment_data/file/61152/Government-Construction-Strategy_0.pdf

Chegu Badrinath, A., Chang, Y. T., & Hsieh, S. H. (2016). A review of tertiary BIM education for advanced engineering communication with visualisation. *Visualisation in Engineering*, *4*(1), 9. doi:10.118640327-016-0038-6

Deutsch, R. (2011). *BIM and Integrated Design: Strategies for Architectural Practice*. John Wiley & Sons, Inc.

Feagin, J., Orum, A., & Sjoberg, G. (Eds.). (1991). *A case for case study*. Chapel Hill, NC: University of North Carolina Press.

Gerber, D., Khashe, S., & Smith, I. (2013). Surveying the Evolution of Computing in Architecture, Engineering, and Construction Education. *Journal of Computing in Civil Engineering*, (Oct): 7.

Gledson, B., Hilton, D., & Rogage, K. (2016). Benchmarking BIM Levels of Training and Education amongst Construction Management Practitioners. *Proceedings of the 32nd Annual ARCOM Conference*.

Gordon, C., Azambuja, M., & Werner, A. M. (2009). BIM across the construction curriculum. *Proc. of the 2009 ASC Region III Conference*.

Grunert O'Brien, J., Millis, B., & Cohen, M. (2009). *The Course Syllabus: A Learning-Centered Approach*. John Wiley and Sons.

Henderson, L., & Jordan, N.L. (2009, Fall). A Modest Proposal for a Transdisciplinary Curriculum for the Design, Construction, Management and Maintenance of Architecture. *Journal of Building Information Modeling*.

Hu, Z. Z., Tian, P. L., Li, S. W., & Zhang, J. P. (2017). BIM-based integrated delivery technologies for intelligent MEP management in the operation and maintenance phase. *Advances in Engineering Software*. doi:10.1016/j.advengsoft.2017.08.007

Kemmis, S., & McTaggart, R. (2000). *The action research planner* (3rd ed.). Deakin University.

Kolb, D. (1994). *Experiential Learning: Experience as the Source of Learning and Development*. Englewood Cliffs, NJ: Prentice-Hall.

Lewin, K. (1948). *Resolving social conflicts: selected papers on group dynamics*. University of Michigan.

Mark, E., Martens, B., & Oxman, R. (2001). The Ideal Computer Curriculum. Architectural Information Management: eCAADe Helsinki, 168-175.

Masters, J. (1995). The history of action research. *Action Research*.

McGraw Hill. (2009). *Smart Market Report: The Business Value of BIM: Getting Building Information Modeling to the Bottom Line*. Bedford, MA: McGraw Hill Construction.

McGraw Hill. (2012). *Smart Market Report: The Business Value of BIM in North America: Multi-Year Trend Analysis and User Ratings (2007-2012)*. Bedford, MA: McGraw Hill Construction.

Nawari, N. O. (2010). Intelligent Design in AEC Education. *ITcon*, *15*, 306–317.

New Jersey Engineers Australia. (2005). *Getting it Right First Time: A Plan to Reverse declining standards in project design documentation within the building and construction industry*. Engineers Australia. Available at: http://codebim.com/wp-content/uploads/2013/06/Getting-It-Right-The-First-Time.pdf

Oja, S., & Smulyan, L. (1989). *Collaborative action research: a developmental approach*. Falmer Press.

Penttila & Elger. (2006). New professional profiles for international collaboration in design and construction. *26th eCAADe Conference Proceedings*.

Pressman, A. (2007). Integrated Practice in Perspective: A New Model for the Architectural Profession. *Architectural Record*. Retrieved from http://archrecord.construction.com/practice/projDelivery/0705proj-3.asp

Pyecha, J. (1988). *A case study of the application of non-categorical special education in two states*. Chapel Hill, NC: Research Triangle Institute.

Scott, L. (2015). The Changing Landscape of Construction Higher Education. *International Journal of Construction Education and Research*, *11*(2), 78–78. doi:10.1080/15578771.2015.1022454

Scott, L. (2016). Embedding Action Research in the Built Environment: An Action Approach. In Research Methodology in the Built Environment. Routledge.

Scott, L., & Fortune, C. (2013). Towards the Improvement of the Student Experience of Assessment and Feedback in Construction Management Education. *European Journal of Engineering Education, 38*(6), 661–670. doi:10.1080/03043797.2013.766675

Scott, L. M. (2016). Theory and research in construction education: The case for pragmatism. *Construction Management and Economics, 34*(7-8), 552–560. doi:10.1080/01446193.2016.1151539

Shelbourn, M., Macdonald, J., McCuen, T., & Lee, S. (2017). Students' perceptions of BIM education in the higher education sector: A UK and US perspective. *Industry and Higher Education, 31*(5), 293–304. doi:10.1177/0950422217725962

Shelbourn, M., Macdonald, J., & Mills, J. E. (2016). Developing an international framework for BIM education in the HE sector. *Proceedings of the 10th Academic Interoperability Coalition (AIC) BIM Symposium.*

Starzyk, G. F., & McDonald, M. (2010). The Collaborative Dance: Only Three Steps. *Proc of the BIM-Related Academic Workshop.*

Topping, K., & Stewart, E. (1998). *Peer Assisted Learning*. Lawrence Erlbaum Associates.

Tran, D., Nguyen, L., & Faught, A. (2017). Examination of communication processes in design-build project delivery in building construction. *Engineering, Construction, and Architectural Management, 24*(6), 1319–1336. doi:10.1108/ECAM-12-2015-0192

Vygotsky, L. (1978). *The Mind in Society – The Development of Higher Psychological Processes*. The Harvard University Press.

Williams, A., Sher, W., Simmons, C., Dosen, A., & Pitt, B. (2009). *Construction Education in Australia: a review of learning and teaching challenges and opportunities*. Australian Learning and Teaching Council.

ADDITIONAL READING

Mahdjoubi, L., & Laing, B. (2015). *Building Information Modelling (BIM) in Design*. Construction and Operations WIT Press.

KEY TERMS AND DEFINITIONS

Active Learning: Active learning is a student-centered approach in which the responsibility for learning is placed upon the student, where they often working in collaboration with other team members. In active learning the role of the teacher/coach is one of facilitation rather than one-way provider of knowledge/information.

BIM: BIM or building information modelling is a process for creating and managing information on a construction project across the project lifecycle.

Case Study: A case study is a research approach involving an up-close, in-depth, and detailed examination of a subject of study (the case), as well as any related contextual conditions.

Collaborative Learning: Collaborative learning is an umbrella term for a variety of educational/ pedagogical approaches involving joint intellectual effort by students with students, or students and teachers together. Collaborative learning has been practiced and studied since the early 1900s.The principles are based on the theories of the greater educationalists John Dewey, Lev Vygotsky, and Benjamin Bloom.

Collaborative Pedagogy: The collaborative pedagogy approach holds true to the position that students will better engage with critical thinking, thinking and revision if they engage with each other.

Design Build: Design-build, in construction terms, is a method of project delivery in which one entity—the design-build team—works under a single contract with the project owner (the client) to provide design and construction services. The Client manages only one contract with a single point of responsibility. The designer and contractor work together from the beginning, as a team, providing unified project recommendations to fit the owner's schedule and budget.

Experiential Research: Experiential learning is a process of learning where the learner is exposed to experience. "A learning by doing" approach that is more specifically defined as learning through reflection on doing. Reflection is an important part of the experiential learning process.

Student-Centered Learning: Student-centered learning is sometimes referred to as learner-centered education. The approach broadly encompasses methods of teaching that shift the focus of instruction from the teacher to the student. This type of pedagogical approach believes that the student should be the focus of the learning and that should be placed at the center of the process.

Chapter 5
Re-Purposing Summative Assessment as Formative:
A Reflective Guide to Facilitating Deep Learning

Obuks Augustine Ejohwomu
The University of Manchester, UK

ABSTRACT

Students often feel that assessments are used primarily for grading purposes only. There is anecdotal evidence too that students are being over assessed in UK Higher Education (HE) given that it is common place for the curricula to be designed with formative and summative assessments in mind. This reflective case study aims to establish the link between the attainment of independent learning outcome and a new form of assessment that is based on three years of teaching a Project Management unit in two internationally leading Schools of Mechanical, Aerospace, and Civil Engineering in England and Singapore. Data from the observation of an operational unit specification reveals that depending on the timing within the schedule of the unit, summative assessment with appropriate feedback can be re-purposed as formative and summative. The implication being a paradigm shift in the design of unit specification for the purpose of attaining Intended Learning Outcomes (ILO) that enhance deep learning.

DOI: 10.4018/978-1-5225-8452-0.ch005

INTRODUCTION

The design of most curricula consists of both formative and summative assessments (Taras 2005). It is argued that the former focuses on the process and the latter on the product (Black et al 2004:55-56). The implication here is emerging evidence of discontentment amongst students and teachers. The students complain of over assessment while few teachers are motivated to operate parallel assessment that is designed for both summative and formative functions (William 2000b:16).

It is therefore not a surprise that the UK Professional Standards Framework aims to ascribe values and ethics with respect for learners and diverse learning communities, promoting participation and equality of opportunity for learners and use of evidence-informed approaches from research and scholarship (UKPSF 2011). Similarly, the affective domain in Bloom's Taxonomy (Krathwohl *et al* 1964) is about learning judgment, discretion and other personal attributes that might be regarded as part of ethics. However, there are not many generalisable quality standards for assessing academic practice and quality (Martensson *et al* 2016) - or more importantly the impact of teaching on the satisfaction of learners. There is also evidence that explicit approach to academic practice can produce a negative response in students who not only find the subject a difficult one but also can regard a didactic teaching approach as ineffective (MacFarlane 2004).

The author will argue that value, quality and ethics are important to academic practice (teaching, research and service). Svinicki (1994) suggests the HE sector seek ethical assessment responsibility not only towards students but also towards training providers.

It is against this backdrop that this book chapter aims to establish the link between the attainment of Intended Learning Outcomes (ILO) and a peculiar (re-purposing of summative assessment as formative) form of assessment that draws on several years of teaching a project management unit in two world leading Schools of Mechanical, Aerospace and Civil Engineering in UK and Singapore.

The chapter will start with a brief literature review and evaluation of a curriculum design from an employability perspective in the Higher Education (HE) sector and a push for wholesome academic practice. This will be followed with a reflective position on assessment and feedback that the author teaches to postgraduate students.

In particular, the chapter will evaluate the ILO (without appending the actual unit specification because of ethical issues), delivery methods, content and some materials used in the case study. The chapter will also anonymously reveal the experiential development of a substantial summative assessment for a credit-bearing unit. This will reflect how it is aligned to the course unit's ILO, its corresponding assessment rubric and constructive feedbacks.

Lastly, the chapter's argument on teaching, assessment and feedback will be supported with some theoretical discussions as well as pedagogic literature. On the whole, the objective of this chapter is in the assertion that depending on the timing within the schedule of the unit, summative assessment with appropriate feedback can be re-purposed as a formative assessment that encourages student satisfaction by way of less is more concept of assessment.

Formative and Summative Assessment

The role of formative and summative assessments in academic practices could be clearer to students and teachers. Commentators with divergent and convergent views have tried to address this from different lenses (see Scriven, 1967; Ramaprasad, 1983; Sadler, 1989; Black *et al,* 2004; Taras, 2005; Biggs & Tang, 2011). For example, Taras (2005) argues that all assessments begin with summative and that formative assessment is in fact summative plus feedback. Harden and James (1997) prescribed alternative procedure for linking formative and summative functions. Others have argued for the re-purposing of summative assessment as formative (Scriven, 1967; Ramaprasad, 1983; Sadler, 1989); Biggs and Tang (2011:195) argue that both formative and summative functions aim to see how well students are doing but that the purpose of the two forms of assessment are divergent. This chapter contribution which is in alignment with the call to re-purpose summative assessment as formative is a new contribution to the limited pool of studies that demonstrate the practical operationalisation of this paradigm shift.

METHOD

The preceding section discussed the repurposing of summative assessment. The objective here is to demonstrate how this push for paradigm shift can be operationalised in practice by way of a reflective case study (see Schon, 1983; Piekkari *et al,* 2009). This is a practical reflection on three years of heavy teaching and assessment that led to the award of Fellow of the Higher Education Academy (FHEA) in recognition of attainment against the UK professional standards framework. The research laboratory (data source) included curricula design, facilitation, and alignment of assessment and feedback in order to achieve ILO.

The average class size in the reflective case study tends to range between 5 - 20 delegates in the small group and 250 – 360 students in the large group. What the author found common to both groups is that quite a considerable amount of time is spent reflecting pre-action, in-action and on-action. Table 1 shows some of the typical questions the author considers when reflecting.

Table 1. Reflective questions

	Pre-Action	On-Action	In-Action
1	Am I using the appropriate pedagogical strategy for my set ILOs?	Was the teaching and learning context suitable for the level and stage of my students from a developmental perspective	Do I want to stop the on-going interaction because of allotted time?
2	Am I well resourced in readiness to teach?	What went did not go well for continuous improvement?	Will the students sustain this deep level of interaction if it were a discussion board activity?
3	Is my content suitable for my cohort and characteristics of my discipline?	Will a flipped class have enlivened the interaction more?	Why is the engagement different from the previous?

(adapted from Gore and Zeichner 1991)

CRITICAL REVIEW OF AN OPERATIONAL COURSE UNIT SPECIFICATION FROM AN EMPLOYABILITY PERSPECTIVE

The unit specification, as the basic building block of any programme should at a first glance relay the unit's ILOs. Where ILOs "…are statements written from the student's perspective, indicating the level of understanding and performance they are expected to achieve as a result of engaging in the teaching and learning experience" (Biggs & Tang, 2011:100-111). The ILO's design will normally be based on the constructivist theory and/or alignment, which is a principle in curriculum theory. A unit specification can therefore be designed for a programme (e.g. MSc in Management of Projects) or for a Unit e.g. MACE 2XX1 – this is a second year Mechanical and Aerospace unit with a 10 credit loading. See Biggs and Tang (2011) for a detailed guide on curricula design for attainment of ILO.

Adopted Critical Review Framework

Employability here is defined as a combination of skills, understanding and personal attributes (Yorke 2006) and a development of critical and reflective abilities (Harvey 2003). This definition, which is adopted in this contribution, is a hybrid of two leading schools of thoughts on what constitutes employability. The first school limits its definition to teaching and learning that promote employability for the student (captured largely as Tier 1 in Figure 1) and the second focuses on effective lifelong learning through critical and reflective thinking (captured largely Tier 2 in Figure 1).

Embedded clearly in Figure 1, too, are the characteristics of Harvey (2003) and Yorke (2006)'s definition of employability. The former can also be perceived to promote the acquisition of declarative knowledge and the latter favours more

Figure 1. A practical model of employability
Adopted from Dacre Pool & Sewell (2007)

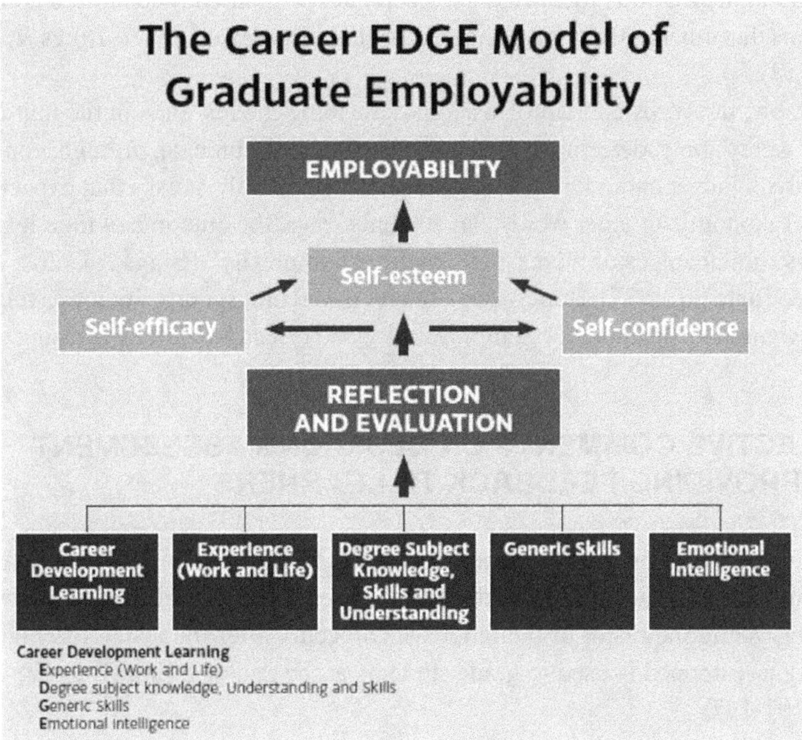

of functional knowledge (see Biggs & Tang pp.92). Even though it is outside the scope of the chapter contribution it is worth noting that some critiques of the implementation of the employment model captured in Figure 1 argue that it has serious resource implications for the teacher and professional identity issues for the student (see Jackson, 2016;Rowe & Zegward, 2017). The lesson here is the need for periodic reflective review of unit specification.

Case Study: A Critique of an Operational Unit Specification

The assessment of the unit specification revealed ten aims. This is rather too many. It can be likened to a dissertation having ten aims. The unit specification did not surprisingly state the ILO that the unit set out to achieve. This was completely empty even though there was provision in the template for it to be inserted. One would have expected that both *Threshold and Core Concepts* of the unit to feature conspicuously (see Meyer & Land, 2003). The likely reason for missing out on the core concepts may be having too many aims. The implication here is that neither tier's of skills identified in Figure 1 is imparted after operationalizing the unit specification. It is safe to argue that this identified gap correlates with already identified common problems that emerge from the designing of ILOs (see Ewell, 2007; Biggs & Tang, 2011:117).

Notably, the verbs and words used in describing the ten aims in the unit under review were "apply, determine, evaluate, specify, communicate, through, consider, summarise, judge, and address". These do not completely satisfy that expected of SOLO Taxonomy. In other words, "as students learn, the outcome of their learning displays similar stages of increasing structural complexity" (Biggs & Collis, 1982; Biggs & Tang, 2011: 87). Simply put – the reviewed unit specification was found to be deficient in its promotion of employability and effective lifelong learning skills.

REFLECTIVE COMMENTS ON DESIGNING ASSESSMENT AND PROVIDING FEEDBACK TO LEARNERS

Whether formative or summative an assessment should aim to achieve set ILO in any unit. The difference being that the result of the former is used as feedback during learning. While the result of the latter which occurs after the actual teaching and learning has stopped is used to grade students at the end of a Unit (Biggs & Tang, 2011: 191-197).

This assessment and feedback principle, which draws on the teachers and student's perspective framework shown in Figure 2 is driven by attainment of ILO. This is because the framework helps with the management of teacher and student expectations – a recipe for a functional wellbeing management system.

When preparing for examination students tend to focus more on studying aspects of the curricula that they suspect will feature in their examination paper. Questions

Figure 2. Teachers and students' perspective
(adapted from Biggs and Tang 2011)

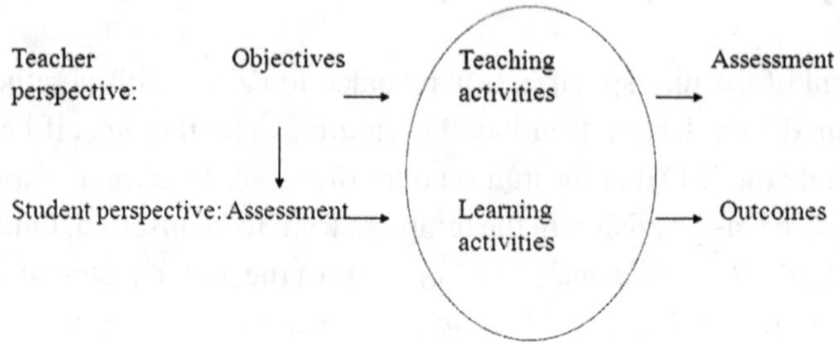

like "will this form part of the exam" are now commonplaces in lecture theatres. This phenomenon termed 'back wash' by Elton (1987: 92) encourages more surface learning than deep learning. Figure 2 is suggesting that teachers could develop a deep learning strategy that encourages positive back wash by embedding ILO (captured as objective in Figure 2) in their assessments. The arrows connote linkages while the circle signals seamless embedding of assessment and ILO through engaging teaching and learning activities. Gamification and facilitation (not teaching) are examples of strategies which the author has successfully used to encourage positive back wash.

Reflection on the Development of a Substantial Summative Assessment

This unit's summative assessment (see Appendix 2) is designed for delegates on the Project Management Professional Development Programme. The average delegate has a minimum of 5-10 years professional experience. To this end, it is better to make the reason for the assessment design explicit and meaningfully relevant (real life industry based problem). Thus, minimising issues of backwash and surface approach to learning (see Biggs & Tang 2011: 24-27; and 191-197).

Although this is a summative assessment, in the experience the author (unit coordinator), it is hard to divorce it completely from formative assessment. In other words, the author will argue that formative and summative are not mutually exclusive. Consequently, delegates mid-term presentation which has a weighting of 20% continuous assessment allows for delegate's 'confident learning' from peers and individual mistakes. In turn, the unit coordinator will issue a personalised feedback (see Appendix 2) detailing positives and signposts areas for improvement. This approach when compared with the control experiment (a unit without formative assessment) the author has discovered that the learning outcome in better satisfied when a formative assessment is built into the supposedly summative assessment. It is worth cautioning here that this approach may not be very suitable for class sizes that exceed forty-five delegates.

RECOMMENDATIONS

1. The unit lead needs to ensure the learning outcomes in this case aims better aligns with the five levels of SOLO Taxonomy. This will in turn encourage the design of a unit specification that promotes employability skills that are embedded in deep learning and lifelong learning outcomes.
2. The unit lead needs to consider the introduction of "graduate podcast for employability" (see Pegs et al 2006 pg 31). In turn, this will allow for good

practice and application of the reflection and evaluation signposted in the adopted review's framework.

3. The number of aims should be reduced from ten to a maximum of six and it should state very clearly what the threshold and core concepts are.
4. The need for paradigm shift by way of re-purposing of summative assessment as both formative and summative cannot be overemphasised. The structure of the assessment and timing of the feedback is fundamental to the effectiveness of this prescribed approach (see Appendix 2).

FUTURE RESEARCH DIRECTIONS

With the spike in mental health cases, HE sectors have started paying closer attention to the monitoring of student well being. Some leading universities have revitalised their mentoring systems such that adequate training is in place to help academics to better identify and respond to the symptoms of poor wellbeing. While this can be one way of managing the crisis this chapter is calling for more pedagogic studies into the correlation between assessment and wellbeing. Academics are scrutinised when student feedback is poor, which may have implication on their career progression. Similarly, students feel somewhat dejected when employers tell them they lack the requisite skill to be employed. In other words, any modest improvement to the attainment of ILO will positively impact the wellbeing of academics and students. The main issue that this chapter is trying to address is that academics may need to start rethinking the design or alignment of their unit's specification. By this the author argues that the norm of treating formative and summative assessment as two completely separate piece of assessment may be considered as over assessment of students - an approach, which has implication on student wellbeing and National Student Survey (NSS) results.

CONCLUSION

Academic practice stakeholders believe that assessment is an integral part of the student experience in any HE. But students particularly feel assessments are used for the purpose of grading only. This three year reflective case study tried to establish the link between that attainment of ILO and a new form of assessment. Data from the observation of an operational unit specification reveals that depending on the timing within the schedule of the unit, summative assessment with appropriate feedback can be timely re-purposed as formative and summative. The implication is a paradigm shift in the design of unit specification for the purpose of attaining ILO that enhances deep learning.

REFERENCES

Biggs, J. B. (2011). *Teaching for quality learning at university: What the student does*. McGraw-Hill Education.

Black, P. (2003c). *Formative and summative assessment: Can they serve learning together?* Paper presented at AERA, SIG Classroom Assessment Meeting.

Black, P., Harrison, C., Lee, C., Marshall, B., & William, D. (2004). *Assessment for Learning. Putting it into Practice*. Maidenhead, UK: Open University Press.

Dacre Pool, L., & Sewell, P. (2007). The key to employability: Developing a practical model of graduate employability. *Education + Training*, *49*(4), 277–289. doi:10.1108/00400910710754435

Ewell, P. (2007). *Applying learning outcomes concepts to higher education: An overview*. Prepared for the Hong Kong University Grants Committee.

Gore, J., & Zeichner, K. (1991). Action research and reflective teaching in preservice teacher education – a case study from the United States. *Teaching and Teacher Education*, *7*(2), 119–136. doi:10.1016/0742-051X(91)90022-H

Harlen, W., & James, M. (1997). Assessment and learning: Differences and relationships between formative and summative assessment. *Assessment in Education: Principles, Policy & Practice*, *4*(3), 365–379. doi:10.1080/0969594970040304

Harvey, L. (2003). *Transitions from higher education to work: a briefing paper prepared by Lee Harvey (Centre for Research and Evaluation, Sheffield Hallam University), with advice from ESECT and LTSN Generic Centre colleagues*. Sheffield Hallum University.

Jackson, D. (2016). Re-conceptualising graduate employability: The importance of pre-professional identity. *Higher Education Research & Development*, *35*(5), 925–939. doi:10.1080/07294360.2016.1139551

Meyer, J., & Land, R. (2003). *Threshold concepts and troublesome knowledge: Linkages to ways of thinking and practicing within the disciplines*. Edinburgh, UK: University of Edinburgh.

Pegg, A., Waldock, J., Hendy-Isaac, S., & Lawton, R. (2012). *Pedagogy for employability. Pedagogy For Employability*. York, UK: Higher Education Academy.

Piekkari, R., Welch, C., & Paavilainen, E. (2009). The Case Study as Disciplinary Convention. *Organizational Research Methods*, *12*(3), 567–589. doi:10.1177/1094428108319905

Ramaprasad, A. (1983). On the definition of feedback. *Behavioral Science*, *28*(1), 4–13. doi:10.1002/bs.3830280103

Rowe, A.D., & Zegwaard, K.E. (2017). *Developing graduate employability skills and attributes: Curriculum enhancement through work-integrated learning*. Academic Press.

Sadler, D. R. (1989). Formative assessment and the design of instructional systems. *Instructional Science*, *18*(2), 119–144. doi:10.1007/BF00117714

Schon, D. A. (1983). *The Reflective Practitioner: how professionals think in action*. New York: Basic Books.

Scriven, M. (1967). The methodology of evaluation. In *Perspectives on Curriculum Evaluation (AERA Monograph Series – Curriculum Evaluation)*. Chicago: Rand McNally and Co.

Taras, M. (2005). Assessment–summative and formative–some theoretical reflections. *British Journal of Educational Studies*, *53*(4), 466–478. doi:10.1111/j.1467-8527.2005.00307.x

William, D. (2000b). *Integrating summative and formative functions of assessment*. Keynote address to the European Association for Educational Assessment, Prague, Czech Republic.

Yorke, M. (2006). *Employability in higher education: what it is-what it is not* (Vol. 1). York, UK: Higher Education Academy.

ADDITIONAL READING

Department for Education. (2017) *Teaching Excellence and Student Outcomes Framework: Lessons learned from Year Two*. October Accessed through https://www.gov.uk/government/publications/teaching-excellence-and-student-outcomes-framework- specification on 1 August 2018

Honey, P., & Mumford, A. (2000). *The Learning Styles Helper's Guide*. Maidenhead: Peter Honey Publications Ltd.

McLeish, J. (1968). *The Lecture Method*. Cambridge, UK: Cambridge Institute of Education.

Rutherford, R. J. (1978). Medical student concentration during lectures. *Lancet*, *312*(8088), 514–516. doi:10.1016/S0140-6736(78)92233-X PMID:79879

KEY TERMS AND DEFINITIONS

Formative Assessment: The results of formative assessments are used for feedback during learning.

Summative Assessment: The results of summative assessments are used to grade students at the end of the course or to accredit at the end of a program.

APPENDIX 1

See Table 2.

Table 2.

	Part 1: Substantial Summative Assessment – 1250 words
Tutor:	**Name: Dr. Obuks Ejohwomu** **Email:** obuks.ejohwomu@manchester.ac.uk
Issued:	**Wednesday April 26th 2017**
Requirements:	**Electronic Submission via Turnitin/Blackboard**
Presentation Submission:	Folder open: **Monday July 3rd 2017** Folder closes: **Sunday July 16th 2016 at 23.59 GMT**
Study Session – UoM: Theme 1 & 2	**Wednesday 26th July 2017**
Draft Submission:	Folder open: **Monday August 28th 2017** Folder closes: **Sunday September 10th 2017**
Final Assignment Submission:	Folder open: **Monday September 18th 2017** Folder closes: **Sunday October 8th 2017 23.59 GMT**

Note 1: Presentation of Assignment

The presentation shall be of a professional standard normally expected within the employing organisation.

Note 2: Draft Assignment Submissions

Please note the Module Tutor may be willing to look over **one draft** copy of your assignment prior to official submission and give some general feedback – please email the tutor in advance to make this request. Due to class numbers, please note this is **one draft only per delegate** and the **draft can be submitted for review into a Draft Submission folder in Blackboard (unless otherwise advised by the tutor).**

Note 3: Final Assignment Submissions

Please do not put your name on the assignment – just use your student ID number.

Anonymous marking will be used for all assignments – when you upload your assignment to Blackboard/Tii your name will not be visible to the tutor, so to ensure anonymity please only put your student ID number on the assignment.

Note 4: Request for Mitigation – Assignment Extension Requests / Or Module Deferral (Interruption)

Delegates must choose to either submit a mitigating circumstances form seeking an extension, or a form seeking a deferral / interruption in study. Submitting for both is not an option, so you must decide which route is most appropriate for your circumstances. It is important to make it clear on the form which route you have chosen to apply for.

4.1 Assignment Extension

Delegates who wish to request an extension to the submission of their assignment need to submit a mitigating circumstances request form – this must be submitted as soon as possible but **no later than 2 weeks prior** to the assignment deadline. **Requests received after this date will not be considered.**

4.2 Module Deferral / Interruption

Delegates who wish to request a deferral / interruption to their study on a module need to submit a mitigating circumstances request form – this must be submitted as soon as possible but **no later than 2 weeks prior** to the end of the study cycle.

The guidelines on what does and does not constitute mitigation, and a copy of the relevant form(s) for submitting a request for mitigation (assignment extension or module deferral/interruption) can be found within the course admin area within the **PM PDP Community Unit in Blackboard.**

Note 4: Late Assignment Submissions

Delegates failing to meet their assignment submission deadline – please note your coursework when submitted may be subject to a penalty percentage mark deduction

Note 5: PM PDP Research

The material in the coursework assignment is a valuable source of curriculum development for CC QQM and ideas for research. From time to time the CC QQM Management Team in consultation with the Steering Group may view or give access to the module coursework to external professionals for these purposes, **but** no names or organisations will be disclosed and all material will be made anonymous.

Additionally, please ensure any project that you discuss / reference within your coursework has the relevant company approval. Please note that this includes diagrams, tables and charts.

PART 1

1. **Title**: RISK IN CONDITIONS OF CONTRACT
2. **Aim**: To form part of the MSc examinations.
3. **Objectives:** To demonstrate understanding of concepts through practical application

Background

The questions that form the assignment relate to commercial risk and relationships. This topic is the focus of Module XX plenary and midterm session, the related work required of students prior to the final assignment report submission and draws on many aspects of the module scope and students work experiences.

Assignment

Assignment – Please answer these questions as part of the continuous and final assessment for this module.

Question 1

Explain your current role and experience in commercial and procurement practice as it relates to project management functions, and what benefits you hope to get from this module, in particular which part of the assignment can be most relevant/ helpful to you and why?

(4 marks)

Question 2

Using your firm as case laboratory answer the following questions:

(a. i) From your own experience define "contract cycle time" and estimate its average duration.

(b. ii) How can you reduce contract cycle time in your firm?

(c.) Conceive a "contract scenario" in your firm and set out a contract that will keep Total Cost of Operations (TCO) to the barest minimum (TCO).

(d.) Develop a concession matrix for the scenario described in (c).

(e.) Which of the four (4) listed cost model listed below will you base your negotiation on and why?

  a. Must cost
  b. Should cost
  c. Benchmarking
  d. Total cost

(f.) Drawing on your scenario, derive a Win-Win negotiation strategy that gives suppliers high value, low Total Cost of Operation (TCO) concession that achieves your biggest objectives.

(g.) Prescribe a strategy that will reduce cost in your firm's supply chain.

(56 Marks)

Question 3

With the aid of a diagram, map and evaluate the role of formal and informal networks in your firm?

(15 Marks)

Question 4

Identify and describe specific action(s) or knowledge gained from this module that you will take integrate into your job/function/company? Also, discuss the anticipated benefits and outcomes of taking such action(s)?

(5 Marks)

Form of Submission

One overall assignment submission is required encompassing questions 1 to 4 to be submitted at the end of the study cycle.

Feedback: To support the final submission of the assignment, students are required to produce a presentation and submit via Blackboard/Turnitin in accordance with the following:

Q1 – 2 slides that explain:

- the student's role and experience

- benefits of module and assignment relevance

Q2 – 3 slides that explain:

- the defined "contract scenario" with key points to fully provide a context for your analysis
- an overview of the concession matrix

Q3 – 1 slide that provides:

- a mapping of your firm's formal and informal network

The above **feedback** presentation will contribute 20 marks from the marking scheme detailed above towards the student's final overall assessment.

4. **Marking Schema**

Question 1 - 4%
Question 2 - 56%
Question 3 - 15%
Question 4 - 5%
Presentation – 20%

5. Outcome

Your **final module mark** will be issued after the Board of Examiners has verified all assessment (coursework and examination) marks (typically exam boards are in June and November).

Delegates with any questions or concerns over the feedback received should in the first instance speak to the Module Tutor for clarification. Additional marking guidelines can be found in Part 2 below.

6. Word Count

Word count guidance:
Q1 & 2 – 4000 words
Q3 – 1500 words
Q4 – 500 words (associated presentation material and reference documentation are to be included as appendices but do not count towards the word count)

7. **Other information for delegates**: Delegates are reminded to use and cite references as appropriate and include a bibliography.

APPENDIX 2

Additional Feedback on Midterm Presentation: Example 1

Hello XXXX:

Thank you for being part of this Mid-Term interaction. Your presentation was good; and so was your average score – well done!

Below are my thoughts for consideration as you progress work on the final report.

General:

1. You didn't address most of the questions posed heads on. This can be addressed using the same scenario you presented, which was very good.
2. References were missing and I know we had a conversation around this on the day. Please try to draw on journal articles as they attract better weighting.
3. You may benefit from understand and apply the Matrix Concession and models like Kraljic Model Approach in your workbook. Your ability to apply this will be evidence of very good understanding and blending of theory with practice.
4. Slide 3 – what is your understanding of the roles of the commercial manager vs procurement manager? Your understanding this will enable you article your responses to questions asked better.
5. Concession matrix is an invaluable resource tool. Most literature on this will be associated to more of Purchasing and not Procurement. Omid Ghamami has some interesting books and conversations on running purchasing organisations.
6. Slide 7 - - you may want to revisit the presentation of the formal and informal networks. The message is there but it's not a true reflection of reality. I sent some articles earlier on this and you may want follow this think from one the presenters:

Overall - a very good presentation with room for improvement.

Best
Obuks.

Additional Feedback on Midterm Presentation: Example 2

Hello XXX:

Thank you for being part of this Mid-Term interaction. Your presentation was really good and engaging-well done; and so was your average score – well done again!

Below are my thoughts for consideration as you progress work on the final report.

Figure 3.

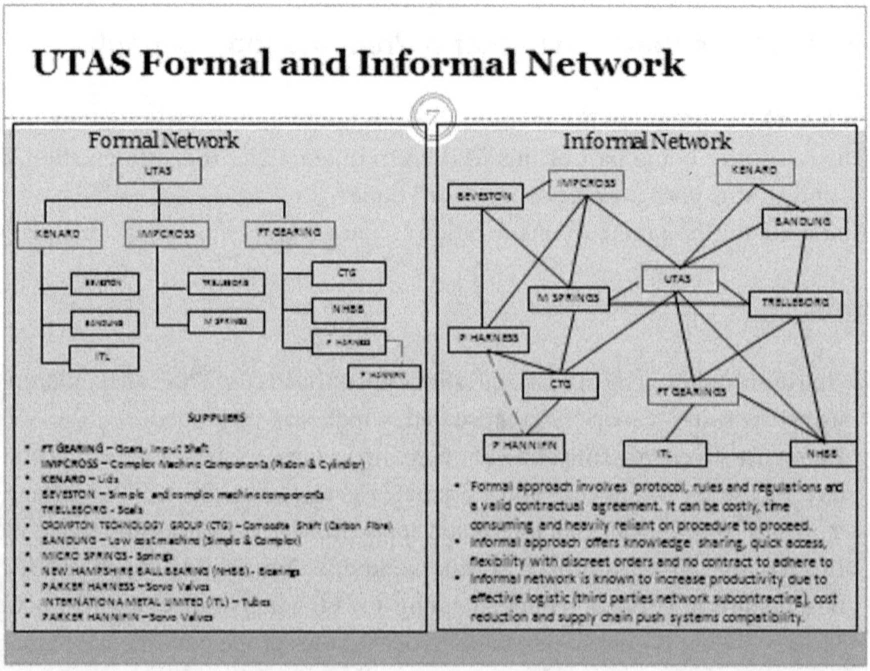

General:

1. Your similarity check was slightly high but within boundaries. The idea here is for you to note this in the preparation of the final report.
2. Slide 4 "..basic principles of lean supply for operational excellence" – what do you mean by this? Will this tend towards Win Win or 50:50? Also, you referenced the multi-stage inventory control systems alongside risk pool, which I think is very interesting. I hope this form part of your discourse. Please use relevant peer reviewed articles to support this.
3. References were missing and I know we had a conversation around this on the day. Please try to draw on journal articles as they attract better weighting.
4. You may benefit from understand and apply the Matrix Concession and models like Kraljic Model Approach in your workbook. Your ability to apply this will be evidence of very good understanding and blending of theory with practice.
5. Slide 5 – there was clear evidence of citation here – well done. However, the format is not exactly correct. Please see BB for links to Harvard referencing style. Still on slide 5 – your follow chart is excellent – well done. However, it will help if you're able to address this question in your final report i). Where/

how is your supply chain interfacing? Can you tie this to your understanding of cycle time in any way supported by references? How is monitoring done and by whom?

6. Slide 6 – How will you define value relative to your organidsation and how this compares to literature definitions? During your presentation you're very clear on the existence of first, second and third their suppliers. If this can be captured in the final report it would be great. In your presentation you referenced a few scenarios that reflected Win Win and not 50:50, which was really impressive. Please ensure this and any other are reflected in your final report.
7. Slide 7 – this was very insightful. Please define what you mean by Need, Want and Now. The logic of how you assigned the ticks would be good to read in your report. The UTC 60 days payment clause can be potentially exploited.
8. Slide 8- this ids a fantastic representation of formal and informal networks. "Discreet order" is very unique and I will hope to see this explained in the final report like you did rightly in the presentation.

Overall, this is a very great presentation with scope for continuous improvement.

Best
Obuks.

Chapter 6
Climate Change in the Built Environment:
Addressing Future Climates in Buildings

Jeremy T. Gibberd
CSIR, South Africa

ABSTRACT

Despite a growing awareness of climate change, there is little evidence that this is being addressed in cities and built environments. Events such as flooding in Houston, USA; landslides in Free Town, Sierra Leone; and water shortages in La Paz, Bolivia and Cape Town in South Africa demonstrate that it is increasingly important that climate change is understood and addressed in built environments to ensure that they become more resilient. This chapter introduces climate change and outlines the implications of this for built environments. It describes measures that can be incorporated into built environments to enable them to adapt to projected climate changes. Understanding climate change and preparing for this by developing built environments that are more resilient will be an increasingly valuable and important skill. Reading this chapter will support the development and refinement of skills and knowledge in this area and it is an essential reference for built environment students and practitioners.

INTRODUCTION

Climate change is now considered as one of the most important issues facing the world today (Hamin & Gurran, 2009). Climate change science has also developed

DOI: 10.4018/978-1-5225-8452-0.ch006

to such as extent that climate change projections can be made within an increasing level of accuracy and detail (Hamin & Gurran, 2009; Guan, 2009).

Simulations now provide climate projections with some degree of certainty 20, 50 and 100 years into the future. Projections describe aspects of climate that will change such as temperatures, the occurrence of very hot days, annual rainfall, heavy rainfall, wind speeds and extreme weather conditions. Current weather data already indicates that climate change is occurring and is confirming the accuracy of climate change projections.

As built environments usually have a lifespan of at least 50 years it is important to understand how climates may change over this period and how this can be addressed in new buildings and urban development proposals (Guan, 2009). Existing built environments must also be assessed in terms of projected climate changes and required adaptations carried out.

This chapter has been developed to support a better understanding of climate change and how it can be addressed in the built environment. It aims to answer the following questions:

- What is climate change?
- What are climate change projections?
- What are the implications of these projections for the built environment?
- How can the built environment be adapted for projected changes?

WHAT IS CLIMATE CHANGE?

Climate change describes a change in the state of the climate that persists for an extended period. This is identified by changes to the mean, for instance, mean temperatures over a long period. Climate change, such as the occurrence of the Ice Ages, can be attributed to natural causes or to human activity. The climate change that is currently being experienced has been directly attributed to human activity (IPCC, 2014).

Current climate change, which is also referred to as global warming, is a result of the greenhouse effect. The greenhouse effect is caused by greenhouse gases, such as carbon dioxide, methane, nitrous oxide and chlorofluorocarbons in the atmosphere trapping heat from the sun and limiting the extent to which this heat is reradiated into space. Increases in these gases in the atmosphere result in a stronger 'greenhouse effect' and leads to higher temperatures and climate change.

Of the greenhouses gases, increases in carbon dioxide have had the most significant impact. Rising carbon dioxide levels are attributed to reduced capacity by vegetation to sequestrate this gas through photosynthesis. It is also attributed to human activities such as the burning of fossil fuels for energy and transport.

Human development and increased urbanization have rapidly reduced areas of natural vegetation, such as forests and grasslands that historically sequestrated carbon dioxide and ensured that there was a long-term balance in the atmospheric gas composition. Increasing generation of energy from fossil fuels as well as the growing use of transportation such as cars and flights powered by fossil fuels has also rapidly increased the amount of carbon dioxide emitted into the atmosphere. As a result, levels of carbon dioxide in the atmosphere have increased from 280 parts per million to over 400 parts per million.

The International Panel on Climate Change (IPCC) warns that the continued increase in greenhouse gases will have very severe consequences, as follows:

Continued emission of greenhouse gases will cause further warming and long-lasting changes in all components of the climate system, increasing the likelihood of severe, pervasive and irreversible impacts for people and ecosystems. Limiting climate change would require substantial and sustained reductions in greenhouse gas emissions which, together with adaptation, can limit climate change risks. (IPCC, 2014)

WHAT ARE CLIMATE CHANGE PROJECTIONS?

Anthropogenic greenhouse gas emissions are related to population size, economic activity, lifestyle, energy use, land use patterns, technology and climate policy. The IPCC has developed Representative Concentration Pathways (RCPs) to make climate change projections, based on these factors. Representative Concentration Pathways (RCPs) are defined according to their contribution to atmospheric radiative forcing in the year 2100 relative to pre-industrial values. An RCP of 8.5, therefore, represents the addition to the earth's radiation budget as a result of an increase in Green House Gases (GHGs) of +8.5 W/m^2. The RCPs used for climate change projections are:

- **RCP2.6:** This represents a stringent mitigation scenario
- **RCP4.5:** This represents an intermediate scenario
- **RCP6.0:** This represents an intermediate scenario
- **RCP8.5:** This represents very limited mitigation and very high GHG emissions

Even though RCP2.6 represents a stringent reduction in carbon dioxide emissions (mitigation), projections indicate that global warming will still occur, but will be below 2°C, relative to pre-industrial temperatures in this scenario (IPCC, 2014).

Other scenarios are projected to result in larger increases in global warming and a range of other impacts. Future climate change, therefore, depends on past anthropomorphic emissions, as well as future emissions and natural climate variability. The following global changes in climate are projected by the IPCC.

Global Mean Surfaces

Global mean surface temperatures will increase by 0.3°C to 0.7°C over the period 2016–2035 relative to 1986–2005. This is similar for all four RCPs.

By 2050 climate change will be markedly different for the different RCPs. Relative to 1850–1900, global surface temperature increases are likely to exceed 1.5°C for RCP4.5 and exceed 2°C for RCP6.0 and RCP8.5.

By 2100, increases in the global mean surface temperature are likely to be 0.3°C to 1.7°C for RCP2.6, 1.1°C to 2.6°C for RCP4.5, 1.4°C to 3.1°C for RCP6.0 and 2.6°C to 4.8°C for RCP8.59.

Climate Change Impacts

Climate change projections indicate that very significant warming is possible in the next 80 years if efforts to mitigate carbon emissions are insufficient. This warming will have a wide range of impacts globally, such as:

- More droughts and heat waves
- Increases in sea levels as ice caps melt
- Changes in growing seasons
- Changes in rainfall amounts and patterns
- Increases in extreme weather conditions such as hurricanes, storms and very heavy rainfall

These impacts will have significant consequences on social and economic development (Müller et al, 2014).

Climate Change Projections for Different Parts of the World

Different areas of the world, however, will be affected differently. Projections are now available for many areas of the world that show the type of changes expected at a local level. These projections are widely available and can be consulted to ascertain changes in your area. As an example, climate change projections for South Africa are provided below.

Example of Climate Change Projections: South Africa

Climate change projections for South Africa have been generated for 2030, 2050 and 2100 for RCP 2.6, RCP, 4.5, RCP 6.0 and RCP 8.5 scenarios. Projections for low mitigation (RCP 8.5) indicate the following broad climatic trends:

- **Hotter Temperatures:** Temperature increases of 1 to 2.5°C in the southern coastal areas and 3°C in the northern areas of South Africa are projected for the period 2021 to 2050, relative to temperatures in the period 1961 – 1990.
- **Minimum Temperatures:** Minimum temperatures are projected to increase by 2 to 3°C for the period 2021 – 2050, relative to the period 1961 -1990.
- **Very Hot Days:** An increase in very hot days is projected for the period 2021 – 2050, relative to 1961 – 1990.
- **Changes in Rainfall:** Increases in annual rainfall are projected in the central interior and east coast, while reductions are expected in the western interior and the north-eastern parts of South Africa in the period 2021-2050, relative to the period 1971 – 2000.
- **Extreme Rainfall Events:** Extreme rainfall events are projected to increase in frequency in the central interior and east coast for the period 2021-2050, relative to the period 1961 – 2000. For the period 2070-2099, relative to the period 1961 – 2000, reductions in these events are projected for Lesotho and Kwa-Zulu Natal Midlands areas.
- **Increased Wind Speeds:** Wind speeds are projected to increase in the northern interior regions of South Africa and decrease in other regions for the period 2021-2050, relative to the period 1961 – 2000 (Engelbrecht, 2017).

A detailed review of these projections indicates that within these broad trends there are local differences. For instance, in some parts of the country, increases in rainfall are expected, while in others, annual rainfall is expected to reduce. It is important therefore to understand not only the range of climate change variability projected under different scenarios but also local differences within a country or area. These variables and their implications for the built environment are explored next.

WHAT ARE THE IMPLICATIONS OF CLIMATE CHANGE PROJECTIONS FOR THE BUILT ENVIRONMENT?

An improving understanding of climate change implies that there should be corresponding improvements in built environment responses. Addressing climate change in buildings requires both mitigation and adaption measures.

Mitigation measures aim to reduce greenhouse gas emissions related to built environments in order to slow down, and ultimately, stop climate change. Adaptation measures aim to ensure that built environments adjust to unavoidable changes that will in occur and minimize their negative impacts (Hamin & Gurran, 2009; IPCC, 2007).

A balanced approach is required, as these measures can be in conflict. For instance, building adaptations to cope with increased wind speeds, storms and floods projected under climate change projections will require additional materials and a heavier structure. This more robust construction results in additional carbon emissions thereby contributing to climate change.

However, synergies are possible. For example, shifting to decentralized renewable energy generation systems reduces both carbon emissions (mitigation) as well as the risk of widespread power outages associated with severe storms (adaptation) (Hamin & Gurran, 2009). Where possible, this type of 'win-win' approach must be pursued.

Although the accuracy of climate change modeling is rapidly improving, it is important to understand that there are still many uncertainties. Measures to address climate change should, therefore, be adaptable and have margins that account for uncertainty (Hallegatte, 2009)

The capacity, or ability, to adapt to future climate change is referred to as resilience. Resilience describes capabilities which enable climate change to be accommodated or adapt to (Hamin & Gurran, 2009). The International Panel on Climate Change describes resilience as the capacity to absorb disturbance and change, while still retaining the same basic structure and functioning (IPCC, 2007). These definitions indicate the need for increased levels of flexibility and adaptability in building system and component design (IPCC, 2007). The science of more resilient built environment design and operation is rapidly developing and is generating a rich palette of tools that designers and managers can use to improve resilience. (Gibberd, 2017a; Gibberd 2017b).

HOW CAN THE BUILT ENVIRONMENT BE ADAPTED FOR PROJECTED CHANGES?

Adapting built environments for climate change should be based on detailed local climate change projections. This data should be used to undertake detailed adaptations which ensure that the built environment is resilient to projected change. In the example below, built environment adaptations which respond to projected climate change in South Africa are provided. These illustrate how built environments can be adapted to respond to climate change.

From the review of climate change projections, the following key changes are envisaged:

- Hotter temperatures
- Increased minimum temperatures
- Very hot days.
- Changes in rainfall
- Extreme rainfall events
- Increased wind speeds

Adaptations to address these changes are explored next.

Higher Average Temperatures

Higher average temperatures will result in temperatures around building increasing. In buildings without insulation or cooling, temperatures are likely to increase. Higher temperatures may mean that many aspects of existing and proposed designs may no longer be suitable. For instance, increased temperatures may result in existing cooling equipment struggling to accommodate higher heat loads. Similarly, insufficient solar control, insulation and passive cooling strategies may also mean that internal conditions in buildings become uncomfortably hot. Higher temperatures will also reduce the lifespan of components and materials that have been designed for lower ambient temperatures.

Higher external temperatures will particularly impact people who work outside such as construction and agricultural workers. It will also affect children and youth who are outside during school breaks and who take part in sports activities.

Addressing this type of climate change requires an approach that is responsive to the local conditions and the site. This approach can draw on the following broad measures and tailor these to suit local circumstances:

- **External Hard Surfaces:** The extent of external hard surfaces, such as roads and paths, on sites should be minimised. Where these exist, surfaces should be lighter coloured. These measures reduce the extent to which heat is absorbed and accumulates around buildings and urban spaces to produce what is known as the urban heat island effect (Zuo et al, 2014; Santamouris, 2015)

- **Trees and Shading:** Shading should be provided around buildings. In particular, this should be provided in external spaces that are occupied during the day, such as school playgrounds. Shading reduces the urban heat island effect and provides cooler external spaces that can be used on hot days (Zuo et al, 2014; Wong & Chen, 2009).
- **Cool Roofs:** Specifying lighter colours on external faces of roofs and walls that are exposed to the sun reduces heat absorption through these surfaces and therefore the heat gain in the building (Cotana et al, 2014; Lapisa et al, 2013). This can reduce the energy consumption required for cooling by 18% (Konopacki et al, 1998). The term albedo is used to describe the "reflectiveness" of surfaces. This captures the extent to which radiation is reflected back, as opposed to what is absorbed in the surface. A low albedo surface has a value nearer 0 and absorbs most of the radiation, while surfaces with a high albedo have a value nearer 1 and reflect most of the radiation.
- **Insulation:** Increasing insulation can be used to reduce heat transfer through the building envelope and can be used to reduce heat gains in the building. Additional insulation in the roof areas of the building envelope exposed to the sun is particularly important in reducing solar heat gains and for improving the effectiveness of passive strategies (Saman et al, 2013).
- **Passive Cooling:** Passive cooling refers to strategies which keep buildings cool with low or no energy consumption and associated carbon emissions. Strategies include cross ventilation, evaporative cooling and night-time cooling. Cross ventilation strategies are suitable for locations with breezes, such as coastal regions. Night-time cooling strategies are suitable for areas with a high diurnal range, such as the interior of the country (Karimpour et al, 2015; Peacock et al, 2010).
- **Mechanical Cooling:** Mechanical cooling systems refer to systems which actively cool buildings through systems that use energy, like air conditioning. Passive systems are preferable to these, but where mechanical systems are necessary, the most energy efficient system should be selected. Ceiling fans, for instance, are much more energy efficient than air conditioning and can be surprisingly effective in many situations. Alternatively, mechanical cooling can be linked to onsite renewable energy generations systems such as a photovoltaic system which has maximum output at the hottest time of the day.

Very Hot Days

Very hot days are defined as days when the maximum temperature exceeds 35°C. An increased number of very hot days (VHDs) have a number of important implications for human health, comfort and the built environment.

In conditioned buildings consecutive very hot days may mean that existing cooling equipment in buildings may not be able to accommodate increased heat loads, leading to uncomfortable conditions in the building. In unconditioned buildings which have insufficient insulation and inadequate passive cooling strategies, internal conditions in buildings will become uncomfortably hot. Very hot conditions over long periods may also affect the lifespan of components and materials that have been designed for lower temperatures.

A range of adaptations can be undertaken to accommodate very hot days. These include personal adaptation and building adaptation measures. The exact nature of these adaptations should be developed to respond to local circumstances.

- **Building Adaptations:** Building adaptations for very hot days aim to keep the building as cool as possible. Adaptations include ensuring that external glazing in buildings have adequate solar protection and are shaded from direct sunshine. A range of interior rooms, courtyards and rooftops that can be moved between to access cooler spaces at different times of the day and night can also be used to provide relief from the heat (Hatvani-Kovacs et al, 2016; Baker & Standeven, 1996). In addition, providing drinking fountains near all occupied spaces enable people to keep cool and well hydrated.
- **Personal Adaptations:** Personal adaptations which reduce heat stress include drinking enough water, spending time indoors or in shaded, cooler areas, wearing loose, light coloured, cotton clothing, and planning any activities to avoid the hottest part of the day (Krecar et al, 2014; Hatvani-Kovacs et al, 2016; Saman et al, 2013)

Minimum Temperatures

Generally warmer conditions resulting from climate change can result in higher minimum temperatures. Implications for buildings in winter include a lower risk of water freezing in pipes and dangerous icy conditions on external surfaces. Increasing minimum temperatures in summer reduce the effectiveness of night-time cooling as a passive environmental control strategy in buildings.

Warmer temperatures in winter also reduce heating loads for mechanical heating systems in buildings and mean that the requirement for heating is reduced or even eliminated in some locations. Indoor temperatures in buildings without mechanical heating will result in improved levels of comfort in winter. Warmer conditions will also mean that performance requirements for aspects such as insulation, glazing, and heating systems may also be less stringent. However, it should be noted that some these aspects, such as insulation, also help to reduce heat gain under hot conditions and therefore should not be reduced as warmer conditions are likely to increase under climate change. The following implications

- **Insulation:** Increasing minimum temperatures may mean that it is easier to achieve comfortable internal conditions during winter. Climate change, however, may also mean that it will be more difficult to keep internal conditions comfortable during summer as higher temperatures are experienced. Therefore, a responsive approach to insulation design and specifications is recommended. This should take into account local climate particularities and passive design strategies including the use of thermal mass. A 'blanket' approach in the application of insulation (which may lead to overheating) should be avoided (Gibberd, 2009b; Porritt et al, 2013; Ren et al, 2014).
- **Heating:** Warmer conditions will reduce heating requirements. However, higher average temperatures may also result in an increased requirement for cooling. This shift will, therefore, need to be addressed in the design and specification of heating and cooling equipment.
- **Passive Design:** Increased minimum temperatures will change the parameters that may have been used in passive design strategies, such as night-time cooling. These strategies there should be revised, based on projected temperatures under climate change.

Changes in Rainfall

Under climate change, some countries, like South Africa, will experience increased rainfall in some areas, while this is reduced in others. Implications for buildings, therefore, need to respond to local projections.

Reductions in rainfall will mean that cities and built environment will be more likely to be affected by water shortages as water supplies from dams, rivers and boreholes are more likely to be depleted. Reduced rainfall will also diminish moisture available for plants and influence the type of plants that will grow leading to the establishment of more drought-resistant species.

More rainfall will increase moisture around buildings. This could lead to increased dampness and moisture-related problems in poorly constructed buildings. If these problems are not addressed, rapid deterioration of the building fabric could occur and associated health problems, such as asthma, could become prevalent (Bornehag et al, 2004; Fisk et al, 2007). Wetter conditions will also affect planting and favour plants that are suited to higher rainfall conditions. The following generic built environment adaptations can be applied to address changes in annual rainfall:

Areas With Decreasing Annual Rainfall

- **Water Efficient Fittings:** Water efficient fittings such as low-flow showerheads and taps should be used to reduce water consumption (Gibberd, 2009a). The installation of baths should be discouraged, and showers should be used instead. Alternatives to flush toilet systems should be investigated and where possible, low water use sanitation systems such as aqua privies and composting toilets specified. Where flush toilets are used, dual flush and low flush toilets should be specified. These measures can be complemented by water metering and leak detection systems which help manage and reduce water consumption.
- **Water Uses:** Water uses that are necessary should be avoided. Examples include domestic swimming pools, ornamental ponds and fountains, car washing and the irrigation of large areas of non-functional landscapes (Gibberd, 2009a).
- **Greywater Systems:** Greywater systems capture water from showers and wash hand basins and other 'cleaner' washing activities and enables this to be reused to flush toilets or for irrigation. These systems should, therefore, be installed in building types where there is a significant production of grey water, such as gyms, student housing and in domestic situations (Gibberd, 2009a).
- **Rainwater Harvesting:** Rainwater harvesting systems capture runoff from roofs and hard surfaces and stores this, enabling it to be available during dry periods. This reduces or even avoids, the need for a mains water supply. This capacity also provides a useful buffer in case of water shortages (Gibberd 2009a; Gibberd 2015).
- **Planting Schemes and Landscaping:** Specification of plants, such as trees, with a higher tolerance for drier conditions, ensure that these survive drier conditions. Landscape management techniques such as mulching and hydro zoning (planting species that have the same water requirements together) also can be applied to enable plants to grow in drier conditions.

Areas With Increasing Annual Rainfall

- **Waterproofing Details:** Increasing moisture levels around buildings can be addressed through high-quality waterproofing in buildings. This includes the correct application damp-proof membrane (DPM) and damp-proof course (DPC). Additional attention should be taken to ensure that roofs and exposed areas of the building are watertight and leaks are avoided. Thus, careful detailing, selection and specification of appropriate components, construction by trained contractors and effective supervision, as well as effective commissioning and testing of completed installations can be applied to ensure that built environments are resilient to increasing annual rainfall.

- **Sustainable Urban Drainage Systems (SUDS):** Sustainable urban drainage systems are designed to reduce the impact of surface water drainage from one site on other sites. This is achieved through measures such as reducing the extent of impervious surfaces such as walkways, roads and parking areas and maximising areas where runoff can be absorbed such as planted areas. In addition, products such as permeable paving which direct runoff to replenish groundwater or be stored and used for irrigation and other purposes. Specific landscaping techniques such as filter strips, swales, and ponds can also be developed to retain and store runoff on site.

Extreme Rainfall Events

Extreme rainfall events are defined as events when 20mm of rain occurs over an area of 64km^2 over 24 hours. Projected increases in extreme rainfall have a number of implications for buildings. Very heavy rainfall on roofs, if not disposed of quickly, can accumulate causing structural collapse. Rapid increases in runoff from hard surfaces can also cause flooding and damage to building in low lying areas.

Built environments can be adapted for these extreme rainfall conditions through the following measures. These should be adapted and design for particular sites and local climate conditions.

- **Location:** Development of low-lying areas that are prone to flooding should be avoided. Climate change modelling can be undertaken to identify these areas and local authorities should ensure that these are not developed.
- **Roof Design and Details:** Current design standards for roofs and rainwater good may not be adequate for projected storm bursts and flows. These should, therefore, be revisited using projected climate change data. Extreme rainfall events can be addressed by ensuring that rainwater goods have a higher capacity and through more numerous outlets, gutters and downpipes. In addition, onsite rainwater storage capacity and infiltration (through SUDs measures) should be re-evaluated to ascertain whether they have sufficient capacity to accommodate projected flows under climate change conditions.
- **Stormwater Design:** Extreme rainfall events will substantially increase volumes of surface runoff and the potential for flooding. This can be addressed by reducing areas that generate runoff (such as parking areas with impervious surfaces) and maximising absorbent areas such as swales.

- **Flooding:** in low lying areas where flooding is possible a number of adaptations to buildings can be considered. Firstly, the ground floor levels can be raised to reduce or avoid flooding the interior of buildings. The entire buildings can also be raised on legs to enable water to flow underneath them avoiding 'damming' that can cause structural collapse. The design and maintenance of stormwater drainage systems should avoid blockages caused by detritus. Blocked drains can rapidly lead to flooding and damage that can easily be avoided if drains are designed and managed appropriately. Equipment and materials that may be damaged by water should be located in higher levels in the building. In addition, more resilient materials should be specified in areas where they may be flooding to reduce impacts such as water damage.

Increased Wind Speeds

Increased wind speeds result in additional stress on exposed elements in buildings such as towers and overhanging sections of the roof. If these elements are not addressed damage from high wind speeds can result. Changes in wind speed also affect energy generation potential in wind turbines. Buildings can be adapted for increased wind speeds in the following ways:

- **Wind Loadings:** Where there will be substantial increases in projected wind speeds guidance on standards and guides on wind loadings in buildings should be reviewed and revised if necessary. In particular, this should address aspects such as wind loadings on roofs, roof overhangs, towers, walls and large areas of glazing. Adaptations to cater for increased wind loading include additional structural and fixing elements.

A STRUCTURED APPROACH TO ADDRESSING CLIMATE CHANGE IN BUILT ENVIRONMENTS

This chapter provides valuable insight and resources on how climate change can be addressed in the built environment. It confirms that a structured and integrated long-term approach based on evidence can be used to ensure that built environments are adapted for climate change and enable key functionality to continue. Key aspects of this structured approach and are summarized below.

Climate Change

The first step in addressing climate change in the built environment is to establish, with as much accuracy as possible, projected climate changes for the local area for the next 20, 50 and 100 years. Projected climate change in the following areas should be obtained: changes in average as well as maximum and minimum temperatures, changes in the number of very hot and very cold days, changes in annual rainfall and variability, changes in extreme rainfall events and changes in extreme weather events including storms and wind speeds.

Implications for the Built Environment

Projected climate change data can then be analyzed in relation to the buildings and built environments that need to be addressed. It is important to understand the functions and vulnerabilities of the proposed or existing buildings. For instance, buildings like hospitals and crèches will be more difficult to evacuate in case of emergencies, such as flooding linked to climate change. In addition, buildings that are in poor conditions, or have been constructed out of less robust materials may be more vulnerable, and less able to withstand higher wind speeds or flooding associated with climate change. Therefore, a key plank in strategies addressing climate change is ensuring that built environments are as inherently as robust as possible, before additional measures to address climate change are added.

Designing for Climate Change

Once climate change projections are fully understood and the existing vulnerabilities of proposed designs, or existing built environments are known, specific interventions to address climate change can be proposed. Designs should draw on standards and codes that take into account future climate change projections, for instance, expected temperatures and rainfall and design for these conditions. If updated codes and standards (that reflect projected climate change) do not exist for the local area, an alternative is consulting codes and standards for an area that is as similar as possible to the projected climate. For example, if an area is projected to become drier in future, best practice codes and standards for the most relevant existing drier area can be drawn on as guidance and refined for local circumstances.

Integration

As the understanding of climate change and building adaptation and mitigation develops, it is important that optimum approaches are captured in standards and guidance to ensure that effective practice is disseminated and shared. This can be done through voluntary and professional associations but should also be incorporated into formal policy and legislation such as building regulations and town planning schemes.

Enhanced Capacity and Systems

In addition to following a structured approach to addressing climate change enhanced capacity and systems will be required. These are described next.

Human Resources

A strong understanding of climate, climate change and building physics is required to begin to understand how built environments can be planned, designed, constructed and operated to become more resilient to climate change. Skills in these areas will become increasingly valuable and should be catered for at higher education institutions and by the relevant built environment professional bodies within the fields of Architecture, Engineering, Landscape Architecture and Quantity Surveying.

Modeling and Simulation

Identifying appropriate built environment responses to projected climate change can be enhanced through modeling and simulation. The software in this area is becoming increasingly powerful and accessible. Simulations can be run for a wide range of scenarios and scales, enabling a wide range of options to be evaluated. This process can be extremely in rapidly identifying optimal solutions and refining these.

Risk

Climate change represents a wide range of changing risks, many of which are not fully understood or calibrated yet. Managing these risks requires a sophisticated understanding of how risk can be managed and addressed in built environments. An increasing awareness of these risks is resulting in specialized roles being created within large organisations and cities to specifically understand and address climate change risks.

CONCLUSION

The impacts of climate change are already apparent. Increasingly sophisticated modeling now enables detailed climate change projections to be developed at a local scale for up to 100 years in the future. Aspects of climate that can be projected include higher temperatures, the occurrence of very hot days, rainfall, extreme rainfall events and wind speeds.

Given the increasing accuracy and detail of these projections, it is now possible to understand their implications for the built environment. Analysis of implications indicates that there will be significant impacts on built environments and that large-scale adaptations are required in order for the built environment to accommodate projected climate change.

Once projections are understood adaptations can be integrated into built environments enabling them to become more resilient, and to adapt more easily to future climates and associated impacts. A detailed list of measures is provided which can be applied to built environments in order to adapt to climate change.

Understanding climate change and preparing for this by developing built environments that are more resilient will be an increasingly valuable and important skill. This chapter provides the basis for the development of skills and knowledge in this area and is essential reading for built environment students and practitioners.

ACKNOWLEDGMENT

This chapter is partly based on climate change and built environment research funded by the National Research Foundation (NRF).

REFERENCES

Baker, N., & Standeven, M. (1996). Thermal comfort for free-running buildings. *Energy and Building*, *23*(3), 175–182. doi:10.1016/0378-7788(95)00942-6

Bornehag, C. G., Sundell, J., Bonini, S., Custovic, A., Malmberg, P., Skerfving, S., ... Verhoeff, A. (2004). Dampness in buildings as a risk factor for health effects, EUROEXPO: A multidisciplinary review of the literature (1998–2000) on dampness and mite exposure in buildings and health effects. *Indoor Air*, *14*(4), 243–257. doi:10.1111/j.1600-0668.2004.00240.x PMID:15217478

Cotana, F., Rossi, F., Filipponi, M., Coccia, V., Pisello, A. L., Bonamente, E., ... Cavalaglio, G. (2014). Albedo control as an effective strategy to tackle Global Warming: A case study. *Applied Energy*, *130*, 641–647. doi:10.1016/j.apenergy.2014.02.065

Engelbrecht, F., (2017). *Detailed projections of future climate change over South Africa*. CSIR Technical Report.

Fisk, W. J., Lei-Gomez, Q., & Mendell, M. J. (2007). Meta-analyses of the associations of respiratory health effects with dampness and mold in homes. *Indoor Air*, *17*(4), 284–296. doi:10.1111/j.1600-0668.2007.00475.x PMID:17661925

Gibberd, J. (2009a). Water Conservation. In Green Building Handbook (vol. 1). Alive 2 Green Publishers.

Gibberd, J. (2009b). Building Envelope. In Green Building Handbook (vol. 1). Alive 2 Green Publishers.

Gibberd, J., (2015). Rainwater harvesting playing a valuable role in increasing the resilience and sustainability of water supply. In *Sustainable Water Resource Handbook* (vol. 6). Alive 2 Green Publishers.

Gibberd, J. (2017a). Water Resilience in Urban Areas. *11th Built Environment Conference*, Durban, South Africa.

Gibberd, J. (2017b). Social Resilience in Urban Areas. *11th Built Environment Conference*, Durban, South Africa.

Guan, L. (2009). Preparation of future weather data to study the impact of climate change on buildings. *Building and Environment*, *44*(4), 793–800. doi:10.1016/j.buildenv.2008.05.021

Hallegatte, S. (2009). Strategies to adapt to an uncertain climate change. *Global Environmental Change*, *19*(2), 240–247. doi:10.1016/j.gloenvcha.2008.12.003

Hamin, E. M., & Gurran, N. (2009). Urban form and climate change: Balancing adaptation and mitigation in the US and Australia. *Habitat International*, *33*(3), 238–245. doi:10.1016/j.habitatint.2008.10.005

Hatvani-Kovacs, G., Belusko, M., Skinner, N., Pockett, J., & Boland, J. (2016). Heat stress risk and resilience in the urban environment. *Sustainable Cities and Society*, *26*, 278–288. doi:10.1016/j.scs.2016.06.019

Hewitson, B. C., & Crane, R. G. (2006). Consensus between GCM climate change projections with empirical downscaling: Precipitation downscaling over South Africa. *International Journal of Climatology*, *26*(10), 1315–1337. doi:10.1002/joc.1314

IPCC. (2007). *Climate change 2007: Synthesis report, fourth assessment report.* Cambridge, UK: International Panel on Climate Change and Cambridge University Press.

Karimpour, M., Belusko, M., Xing, K., Boland, J., & Bruno, F. (2015). Impact of climate change on the design of energy efficient residential building envelopes. *Energy and Building*, *87*, 142–154. doi:10.1016/j.enbuild.2014.10.064

Konopacki, S., Gartland, L., Akbari, H., & Rainer, L. (1998). *Demonstration of energy savings of cool roofs. Executive summary (No. LBNL-40673-Exec. Summ.). Lawrence Berkeley National Lab.* Environmental Energy Technologies Div. doi:10.2172/296884

Krecar, I. M., Kolega, M., & Kunac, S. F. (2014). The Effects of Drinking Water on Attention. *Procedia: Social and Behavioral Sciences*, *159*, 577–583. doi:10.1016/j.sbspro.2014.12.428

Lapisa, R., Bozonnet, E., Abadie, M. O., & Salagnac, P. (2013). Cool roof and ventilation efficiency as passive cooling strategies for commercial low-rise buildings–ground thermal inertia impact. *Advances in Building Energy Research*, *7*(2), 192–208. doi:10.1080/17512549.2013.865559

Müller, C., Waha, K., Bondeau, A., & Heinke, J. (2014). Hotspots of climate change impacts in sub-Saharan Africa and implications for adaptation and development. *Global Change Biology*, *20*(8), 2505–2517. doi:10.1111/gcb.12586 PMID:24796720

Peacock, A. D., Jenkins, D. P., & Kane, D. (2010). Investigating the potential of overheating in UK dwellings as a consequence of extant climate change. *Energy Policy*, *38*(7), 3277–3288. doi:10.1016/j.enpol.2010.01.021

Porritt, S., & Cropper, P., Shao, L., & Goodier, C. (2013). Heat wave adaptations for UK dwellings and development of a retrofit toolkit. *International Journal of Disaster Resilience in the Built Environment*, *4*(3), 269–286. doi:10.1108/IJDRBE-08-2012-0026

Ren, Z., Wang, X., & Chen, D. (2014). Heat stress within energy efficient dwellings in Australia. *Architectural Science Review*, *57*(3), 227–236. doi:10.1080/00038628.2014.903568

Saman, W., Boland, J., Pullen, S., de Dear, R. J., Soebarto, V., Miller, W. F., ... Pockett, J. (2013). *A framework for adaptation of Australian households to heat waves*. National Climate Change Adaptation Research Facility.

Santamouris, M. (2014). On the energy impact of urban heat island and global warming on buildings. *Energy and Building*, *82*, 100–113. doi:10.1016/j.enbuild.2014.07.022

Wong, N. H., & Chen, Y. (2008). *Tropical urban heat islands: climate, buildings and greenery*. Routledge. doi:10.4324/9780203931295

Zuo, J., Pullen, S., Palmer, J., Bennetts, H., Chileshe, N., & Ma, T. (2015). Impacts of heat waves and corresponding measures: A review. *Journal of Cleaner Production*, *92*, 1–12. doi:10.1016/j.jclepro.2014.12.078

ADDITIONAL READING

Adger, W. N., Arnell, N. W., & Tompkins, E. L. (2005). Successful adaptation to climate change across scales. *Global Environmental Change*, *15*(2), 77–86. doi:10.1016/j.gloenvcha.2004.12.005

Australian Academy of Science. (2015). *The Science of Climate Change, Questions and Answers*. Retrieved from https://www.science.org.au/files/userfiles/learning/documents/climate-change-r.pdf

Bates, B. (2009). *Climate Change and Water: IPCC technical paper VI*. World Health Organization.

Bosher, L., Carrillo, P., Dainty, A., Glass, J., & Price, A. (2007). Realising a resilient and sustainable built environment: Towards a strategic agenda for the United Kingdom. *Disasters*, *31*(3), 236–255. doi:10.1111/j.1467-7717.2007.01007.x PMID:17714166

Commonwealth of Australia. (2015). *National Climate Resilience and Adaptation Strategy 2015*. Retrieved from http://www4.unfccc.int/nap/Documents%20NAP/National%20Climate%20Resilience%20and%20Adaptation%20Strategy.pdf

Edenhofer, O., Pichs-Madruga, R., Sokona, Y., Seyboth, K., Matschoss, P., Kadner, S., & von Stechow, C. (2011). *IPCC special report on renewable energy sources and climate change mitigation. Prepared By Working Group III of the Intergovernmental Panel on Climate Change*. Cambridge, UK: Cambridge University Press.

Gething, B., & Puckett, K. (2013). *Design for climate change*. RIBA.

Hallegatte, S. (2009). Strategies to adapt to an uncertain climate change. *Global Environmental Change*, *19*(2), 240–247. doi:10.1016/j.gloenvcha.2008.12.003

Holmes, M. J., & Hacker, J. N. (2007). Climate change, thermal comfort and energy: Meeting the design challenges of the 21st century. *Energy and Building*, *39*(7), 802–814. doi:10.1016/j.enbuild.2007.02.009

Huovila, P., Ala-Juusela, M., Melchert, L., Pouffary, S., Cheng, C. C., Urge-Vorsatz, D., . . . Graham, P. (2009). Buildings and Climate Change: Summary for Decision-Makers. Paris, France: The United Nations Environment Programme-Sustainable Buildings & Climate Initiative. Retrieved from http://www.greeningtheblue.org/sites/default/files/Buildings%20and%20climate%20change_0.pdf

Intergovernmental Panel on Climate Change, (2014). Annex II Glossary. *Climate change*.

Lombardi, D. R., Leach, J. M., & Rogers, C. D. F. (2012). *Designing Resilient Cities*. IHS BRE Press.

Mayor of London. (2011). Managing Risk and Increased Resilience: The Mayor's Climate Change Adaptation Strategy. Retrieved from https://www.london.gov.uk/sites/default/files/gla_migrate_files_destination/Adaptation-oct11.pdf

Paavola, J., & Adger, W. N. (2006). Fair adaptation to climate change. *Ecological Economics*, *56*(4), 594–609. doi:10.1016/j.ecolecon.2005.03.015

Pachauri, R. K., Allen, M. R., Barros, V. R., Broome, J., Cramer, W., Christ, R., & Dubash, N. K. (2014). *Climate change 2014: synthesis report. Contribution of Working Groups I, II and III to the fifth assessment report of the Intergovernmental Panel on Climate Change* (p. 151). IPCC.

Roaf, S., Crichton, D., & Nicol, F. (2009). *Adapting buildings and cities for climate change: a 21st century survival guide*. Routledge.

Royal Society. (2018). *Climate change: evidence and causes*. Retrieved from https://royalsociety.org/topics-policy/projects/climate-change-evidence-causes/

The Met Office. (2018) *Climate Science*. Retrieved from https://www.metoffice.gov.uk/climate-guide

US Climate Resilience Tool Kit. (2018). *Building and Structures*. Retrieved from https://toolkit.climate.gov/topics/built-environment/buildings-and-stuctures

USGBC. (2011). Green Building and Climate Change Resilience, Understanding Impacts and Preparing for Changing Conditions. Retrieved from https://www.usgbc.org/Docs/Archive/General/Docs18496.pdf

KEY TERMS AND DEFINITIONS

Adaptation: Adaptation refers to adjustments to respond to climate change.

Adaptive Capacity: The ability of systems, institutions, humans, and other organisms to adjust to climate change.

Albedo: Albedo is a measure of the "reflectiveness." It is a way of quantifying how much radiation is reflected back, as opposed to that absorbed. Low albedo has a value nearer 0 and absorbs most of the radiation, while high albedo has a value nearer 1 and reflects most of the radiation.

Anthropogenic: Anthropogenic refers to man-made impacts, processes, or products.

Biofuels: Renewable fuels derived from biological materials, such as algae and plants that can be regenerated. This distinguishes them from fossil fuels, which are considered nonrenewable. Example of biofuels are ethanol, methanol, and biodiesel.

Carbon Cycle: Describes the process by which living things absorb carbon from the atmosphere, sediments and soil, or food. To complete the cycle, carbon returns to the atmosphere in the form of carbon dioxide or methane by respiration, combustion, or decay.

Carbon Dioxide: The gas that accounts for about 84% of total U.S. greenhouse gas emissions. In the U.S. the largest source of carbon dioxide (98%) emissions is from the combustion of fossil fuels. Combustion can be from mobile (vehicles) or stationary sources (power plants). As energy use increases, so do carbon dioxide emissions.

Carbon Footprint: A carbon footprint describes the total amount of greenhouse gas emissions caused by an organization, event or product.

Carbon Sequestration: Carbon sequestration removes CO2 from the atmosphere. In nature, this occurs during photosynthesis.

Climate: Climate describes the average and variations of weather in a region over long periods of time.

Climate Change: Describes a change in the state of the climate that persists for an extended period. This is identified by changes to the mean, for instance, mean temperatures over a long period.

Climate Resilience: Climate resilience is the capacity absorb stresses and maintain function in the face of external stresses imposed upon it by climate change and (2) adapt, reorganize, and evolve into more desirable configurations that improve the sustainability of the system, leaving it better prepared for future climate change impacts.

Drought: A drought is a period of abnormally dry weather marked by little or no rain. It lasts long enough to cause water shortages for people and natural systems.

Earth's Atmosphere: This is a layer of gases that surround the Earth.

Emissions: Substances released into the air and are measured by their concentrations, or parts per million, in the atmosphere.

Global Warming: Global warming is the warming trend over the past century.

Greenhouse Gas: The main greenhouse gases are water vapour (H2O), carbon dioxide (CO2), methane (CH4), nitrous oxide (N2O), hydrofluorocarbons (HFCs), perfluorocarbons (PFCs), and sulfur hexafluoride (SF6).

Mitigation: Mitigation involves taking actions to reduce greenhouse gas emissions and to enhance sinks, such as carbon sequestration, in order to reduce the extent of global warming.

Renewable Energy: Energy from sources that will renew themselves within our lifetime. Renewable energy sources include wind, sun, water, biomass (vegetation) and geothermal heat.

RPCs: Representative Concentration Pathways are defined according to their contribution to atmospheric radiative forcing in the year 2100 relative to pre-industrial values. An RCP 8.5 therefore represents the addition to the earth's radiation budget as a result of an increase in GHGs of +8.5 W/m^2.

Chapter 7
Sustainable Practices for Built Environment

Joseph Varghese Thanikal
Amity University, India

Anurita Bhatnagar
Amity University, India

Anupam Jain
Amity University, India

Susan George
Amity University, India

ABSTRACT

This chapter is aimed at appraising the reader with the course delivery methodology/ teaching pedagogy practiced at School of Construction at RICS School of Built Environment, Amity University. The chapter highlights the innovative tools and techniques devised by the school for enhancing a students' learning process and ingraining the thoughts for sustainability in construction at nascent stages of construction management. The course is designed to provide the student with an understanding of sustainable construction principles and communicating current sustainability practice to various stakeholder groups.

INTRODUCTION

The term "sustainability" has an important history in development literature. In 1983, the United Nations convened the World Commission on Environment and

DOI: 10.4018/978-1-5225-8452-0.ch007

Development (WCED), known informally by the name of its chair, Gro Harlem Brundtland. The Brundtland Commission's report, Our Common Future (1987), contains one of the most often cited definitions of sustainability: "Sustainable development is development that meets the needs of the present without compromising the ability of future generations to meet their own needs" (World Commission on Environment and Development, 1987). This definition is sector agnostic and an all-encompassing one from which it is necessary to develop contextual definitions for different sectors. The genesis of sustainability as a theme is a result of global awareness that "initiatives on economic development in the enhancement of production must include the protection of the environment besides social justice and the eradication of human challenges" (Alshuwaikhat Habib M, 2016).

The scope of sustainability is frequently described as including three spheres – social, environmental, and economic. To use an accounting metaphor, sustainability in any project must be evaluated according to a "triple bottom line" of social, environmental, and economic responsibility. Sustainability is at once an integrative discipline and a multidisciplinary school of thought; it has statistical, scientific, and humanistic dimensions. With its focus on specific problems and solutions, sustainability suggests place-based and project-based approaches to student learning. Teaching towards sustainability also reminds us that pedagogy is a civic project; there are important ties between classroom and community. (Vanderbilt University, 2018)

For the built environment sector, Tinker and Burt have derived a more appropriate definition for Built Environment stating "those materials and methods used to construct and maintain a structure that meets the needs of the present without compromising the ability of future generations to meet their own needs" (Tinker, 2003). The built environment sector in general and the construction industry is one of the largest consumers of natural resources. With the world population expected to grow by 1.1 billion over the next 15 years, it is expected that some 600 billion square feet of floor space would be added to accommodate this growth (ClimateworksFoundation). The locus of this estimated growth is expected to be in China, India, and North America due the concentration of population growth in these regions. Needless to say construction professionals in these regions are responsible to ensure judicious use of scarce resources so that the carrying capacity of the land is not disrupted. The process of urbanisation is inevitable across the development that is happening and people migrating to developed area looking for more opportunities. It is the rural area which is getting converted to semi urban and then to urban. The rural population has much disciplined way of life with optimum use resources available and the model still exist as sustainable. The word "Exploitation" is to be seriously considered while we talk about urbanisation or sustainable development.

According to UNEP's Global Status Report the buildings sector continues to grow globally, with floor area reaching nearly 235 billion m^2 in 2016 all over the world. It has also been established that the building and construction industry together accounts for 36% of global energy use and 39% of energy-related carbon dioxide (CO_2) emissions (UN Environment and International Energy Agency, 2017). The construction sector globally accounts for one-fifth of the energy consumption (Architecture2030). The Indian construction industry is not far behind, accounting for close to 40% of total energy consumption in the country (Accenturereport). The building sector must play a pivotal role in reducing fossil fuel emissions to meet the climate goals agreed to in Paris (ClimateworksFoundation). India has undergone rapid urbanisation for the last few decades creating tier system in cities. Much is thought on sustainability in the Indian context, the Government has introduced concept of "Smart cities", "Standardisation", but it is important to understand the sustainability quotient applied to smart cities in the Indian context.

Considering the crucial importance of sustainable consumption of scarce natural resources, it is vital that the providers of built environment industry are sensitized towards the importance of conservation and mindful usage (Singh S., 2018).Globally, academic institutions are the recognised seats of knowledge dissemination. It is thus imminent that these institutions, both public and private, need to systemically integrate the principles of sustainability in construction management education and ingrain the thought process of sustainable built environment in the future leaders. In other words, efforts are required to educate the budding professionals in built environment sector and expose them to the importance of sustainability in construction and its contribution in bringing about sustainable development of the nation. The overall intent of any technical education program is to achieve a systemic integration to build connections and sensitize students and managers to think, operate and feel differently regarding the concept of sustainability (Raufflet, 2013). The number of Indian mega cities will increase from the current five (Mumbai, Delhi, Kolkata, Chennai and Bangalore) to seven by the year 2030 (new additions will be Ahmedabad and Hyderabad), when India will have the largest concentration of mega cities in the world (United Nations, Department of Economic and Social Affairs, 2016). This rapid urbanization has created a gap between the demand and supply of urban infrastructure and services among the major large and medium towns of the country (Ahluwalia, 2011). This phenomenon may promote urban sprawl leading to the congestions in the city. There need to be a proportion against demand and supply to promote comfort, liveable and sustainable growth of cities. The chapter emphasis on inclusion of sustainability in the higher education system.

SUSTAINABLE DEVELOPMENT

The universally accepted definition of Sustainable development is defined as 'ability of humanity to make development sustainable to ensure that it meets the needs of the present without compromising the ability of future generation to meet their own needs' (WCED, 1987, 2013).

The three components of sustainability is environment, economy and equity (Berke, 2002; Godschalk, 2004; Munier, 2005). The three components involve the best benefit to each of the individual. The environment refers to the availability of all-natural resources including safe drinking water and pure air to breath is safe guarded for future generation, economy is that, the process by which the these resources are maintained without an economical cost without any negative impact. Equity refers to social dimension of the society wherein each of the individual benefit from sustainable concept. Smart growth another concept added to sustainability of cities. Due to the technological development and unorganised growth within the centre of cities, there have been phenomenon noted in Europe and other countries, people moving outward of cities to breath fresh air. To avoid the sprawled growth, controls where tried to see that the growth within the inner city was sustainable so that the movement of people towards outer city was prevented (Burchell et al., 1999). In other words, land use and transport facility were implemented so that the growth was taking place in proportion at the inner and outer area of cities. The other point of consideration was the existing population looking at their enhanced living comfort, extending all possible exhaustion of natural resources without caring for future generation. "The preservation and enhancement of social and physical *amenities* of communities that support desired activity patterns, safety, lifestyles and aesthetic values" (Bullard, 2007). As cities aim to provide a liveable environment that offers a quality of life to their citizens in order to promote a healthy and safe environment, urban Indian has attracted investment on the back of strong growth but is failing its many of citizens in providing a liveable environment. Across all major quality of life indicators (shown in the Figure 2), India's cities fall well short of delivering even a basic standard of living for their citizens. As the urban population and its income increase, demand for every key service will increase five to sevenfold in cities of every size and type (McKinsey & Company).

Based on the literature study, it is apparent that Sustainable Development aims at achieving sustainability in the dimensions of environment, economy and equity, and clearly states that each development that has to be carried out with a sustainable approach. Though the concept had a vision it was vague in demonstrating the implementation of various initiated strategies and principles in the field of urban development context. Further, the evolution of urban development paradigms introduced various concepts like Smart Growth, Smart City, etc, focusing on achieving sustainability towards urban development. But, based on the study, it is evident that these concepts have similar and overlapping principles with some unique dimensions added over the years due to the various advancements in fields of technology. Also, as these developments were based on regional contexts, there has been an ambiguity about how to define these developments on the global standard level, thus, evidently indicating there's a need to define such developments with a standard outlook so that they can be scaled uniformly. Hence, based on the above analysis, Smart Development as a concept may be defined as an approach that focuses on improving the quality of life of people, that is based primarily on 3 dimensions; environment, economy, and equity. The aim is to protect, conserve and enhance the environment they live in with accessibility to diverse economic opportunities for everyone. And, with the recent introduction of ICT in the approach, the notion is to make the city as well as the people smart.

Genesis of School of Construction

School of Construction, RICS School of Built Environment, Amity University (RICS SBE) - an industry led academic institution was established in 2013, to serve as a modern world class institution, to deliver quality education to students and aspiring professionals and also to serve as a hub for research and development of technical expertise in built environment. (RICS SBE, 2014). India's first and only School of Built Environment RICS SBE provides management education for built environment professionals at the masters (graduate) level. The school caters to its student body through two distinct streams namely School of Construction (SoC) and School of Real estate and Urban Infrastructure (SoRE).

The focus of this paper is the course curricula delivered under the aegis of School of Construction only. Through this paper, the authors seek to fulfil the following objectives:

- Present an overview of the course pedagogy for MBA CPM and CEQS
- Highlight the integration of sustainability/ sustainable development In the course curricula
- Present the challenges in the delivery of the coursework

The course curricula of the MBA programs of School of Construction have been developed in consultation with Indian construction industry. The key industry participants including the providers of different elements of construction value chain were involved in identifying the expected competencies of future construction management professionals. Their inputs were filtered through the academic lenses with the rigorous brainstorming to ensure the following outcomes:

- The recipients gain knowledge about various phases of the project and activities and issues associated with each phase for efficient and effective project delivery
- The recipients gain knowledge and exposure to various skills, tools and techniques related to project management along with practical exposure for being industry ready
- The recipients are exposed to philosophies like Lean Construction, sustainability, stakeholder management and latest construction techniques and software tools like BIM aiding effective project management

As a result, a student after completing the course work would be well versed in principles of sustainability in the construction sector, is able to critically challenge conventional construction organizational practices and capable to formulate effective methods to optimize resource utility at Construction project sites and offices.

MBA Course Curriculum

The course aims to create a future generation of built environment professionals who are both aware and pro-active in advocating the promotion of sustainable practices in the built environment sector. Along with courses relevant for management in general and construction management in particular, course work for sustainability in construction forms a core component of course curricula. Although sustainability is a global goal, its problems and solutions are always importantly situated in local ecologies and communities (Vanderbilt University, 2018). The major courses imparting sustainability/sustainable development education:

1. Construction and Built Environment Sector, Construction Economics, Pre Construction Planning, Construction Materials and Technology (Semester 1)
2. Services of Built Environment (Semester 2)
3. Sustainable practices for built environment, Building Information Modelling, Quality Health and Safety (Semester 3)
4. Procurement and Supply Chain Management (Semester 4)
5. Laws and regulations in Built Environment (Semester 4)

INTEGRATION OF SUSTAINABILITY IN COURSE CURRICULA

To understand the level of integration of sustainability concepts in the course curricula, School of Construction conducted a survey of its current students and alumni to "evaluate the impact of construction management coursework on the perceived understanding of

Figure 1. Teaching pedagogy
Source: Author's work

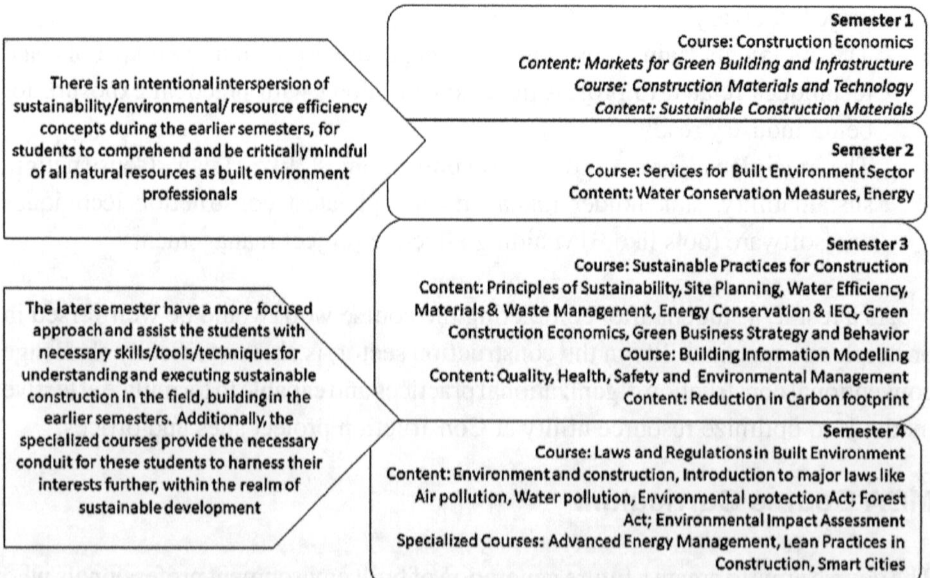

postgraduate students in concepts of sustainability as it applies to the built environment" (Singh S., 2018). The paper concluded that "RICS SBE's curriculum is effective in imparting sustainability concepts in the built environment domain. This can be attributed to two key strengths of the coursework, one being the overall curriculum interspersed with sustainability concepts as opposed to siloed sustainability courses. The second aspect being the transdisciplinary approach towards inclusion of sustainability into the construction management curriculum. RICS SBE has a comprehensive approach of including sustainability concepts into core construction courses during early semesters of the program and then exposing the students to core sustainable construction concepts during the later semesters. This helps not to force fit sustainability but exposes the students towards overall resource efficiency and builds critical thinking as it relates to sustainability concepts" (Singh S., 2018). The figure below represents that overall integration of sustainability course work in the School of Construction's MBA curricula:

Construction and Built Environment Sector, Construction Economics, Pre-Construction Planning, Construction Materials, and Technology

As a part of this course work, the students are introduced to the concept of sustainability through classroom interaction during which they are told about the significance of sustainable development and India's approach towards streamline it in the ongoing initiatives of the government. The course gives an overview of sustainable development in the Indian context and sets the stage for further in-depth studies. The students also spend about 1 week in a semester, learning about measures to incorporate sustainability in pre-feasibility stage (life cycle cost analysis) and value engineering

Sustainable Practices in Construction, Building Information Modelling and Quality Health and Safety

The course, Sustainable practices in Construction, is taught to Semester III students as part of core course work and runs for 15 weeks. The objective of the course is to present sustainability as a key component of the construction industry and how good practices may be translated to achieve sustainability. The course is delivered as a mix of lecture sessions and tutorials wherein the students are appraised of the theoretical aspects of sustainability and its application as seen in case studies. During the structured classroom delivery, the students are made aware of models of sustainability, relevant assessment tools design of sustainable buildings and sustainable building practices. The course aims to encourage students to think of the campus as a sustainability laboratory and emphasizes the important relationship between the university and the larger community of which it is a part.

Technical know-how about the usage of ecological footprints calculators taught during the course and sets the platform for engaging students in important dialogues about wide ranging issues of environmental sustainability and resiliency. The students are exposed to key concepts like ecological footprints for calculations of environmental impacts of the natural resource use. They also gather an understanding of "Carbon Calculators" which are measures of carbon resource consumption and are particularly useful as they pertain to one of the most significant environmental threats and policy issues of our time, namely climate change.

Along with this, the students develop and understanding that the Green buildings demonstrate superior performance than baseline buildings. The subject seeks to introduce students to the techno-managerial aspects of a green building project. Students are exposed to the history and work of international agencies that have led to current important goals such as the Sustainable Development Goals. It seeks to provide to the student an all-round view of sustainability concepts and their relevance and how they can be applied using technology, design, policies, rating systems and proper implementation frameworks in the current context of the built environment. The course enables the student to understand the concepts, models and ideas related to sustainability and the major milestones and organizations involved in the sustainability movement. It provides an understanding of how sustainability is incorporated through policies and the existing government framework for sustainability. The students gain an understanding of green buildings with special emphasis on technology and construction practices, design of the building, sustainable consumption practices and material usage as well as green assessment systems for the built environment. The course also imparts knowledge on application of sustainability in a corporate environment and in greening business operations. Teaching towards sustainability lends itself to place-based and project-based approaches to pedagogy.

Teaching Philosophy

In class the impact of this collective knowledge on the individual and their own arena for bringing change are analyzed. Students are encouraged to translate the teachings into personally relevant areas to produce an evidence of their learning. Current student projects include research papers, blogs, websites, business plans, physical models, and several other creative expressions of their learning. Classroom teaching is supplemented with creative and fun explorations on engagement with high school students and other government and private entities. Teamwork is inculcated through joint exercises that encourage students to apply sustainability concepts outside the classroom. A culmination of a lot of these exercises is the celebration of the World Green Building Week where a School wide celebration is organized each year by students through these activities supplemented by panel discussions, debates, video making competitions, design competition, treasure hunt, green walk, flash mob, street play, competitive quiz, face painting, T-shirt painting, tree planting and many such activities. All these activities are prepped for in the classroom tutorials where international frameworks such as the PRME are used to guide the planning and execution through rigorous discussions and classroom exercises.

Assessments

To ascertain the depth of a student's understanding, the students are assessed in the following manner:

- **Graded Assignments**: To inculcate the habit of self-study, the students are given the opportunity to give them a choice of subject and read on their areas of interest —policies—conferences—energy—water management—ratings and then further work on it—additional commitment towards reading and being able to express they learn
- Following assignments were conducted during the 3rd semester course work:

- **Application Based Case Study**: Study a building and study it w.r.t to LEED pre- certification guide and give the building a rating, recommend how it may be improved (6-7 weeks).
- In class debates/ discussions students take sides and discuss the latest issues relevant to sustainability (development v/s environment, carbon rating etc....) (2-3 weeks).

Procurement and Supply Chain Management and Laws and Regulations

Taught in the IV Semester, this course has sustainability as an inbuilt part of curriculum. Procurement of green materials, creation of green supply chain working with people and systems involved in green supply chain management, green parameters for vendor and contractor selection are some of the course content delivered in the 2 weeklong interaction of a 15 weeklong course. The students are also exposed to the essential components of reverse logistics through structured classroom interactions. In addition to this, the students are encouraged to visit C&D waste recycling unit to understand the concept of reduce-reuse-recycle as well as exposed to drivers of green supply chain via a graded assessment.

Key Initiatives Taken by the School

The school takes initiative in addressing the following issues related to sustainability in built environment: Physical, social, economic and institutional infrastructure. The physical infrastructure includes, services, like, water, sanitation, energy efficiency and management of such services in an optimum way. The design of intelligence system for monitoring and control is part of the education. Social is to address the entertainment and other social gathering places for the proximity of disadvantaged society to be part of development. Economic part involves equal opportunity for investment in infrastructure according to the GDP of the country. Economic way of designing real estate, housing and other possible transport systems can brining in the participation and inclusiveness of society. Institutional part is addressed through the curricula designed to manage the built environment. The education at RICS SBE aim at the possible ways of including direct and indirect courses towards skill development for sustainable development.

CONCLUSION

Sustainable development is being addressed as “Smart” development and many of the developments addressed through coining this word. The smart development cannot be defined with a one unified standardised definition. Smart cities are very commonly used term and tried to practice

Smart City concept focuses on providing the infrastructure services and utilities to urban settlements with the integration of the modern technology (ICT) with the aim to improve the quality of living. The concept has been considered as the future of sustainable development in the field of urban planning but the concept has no unified standards at world level and varies nations to nations.

India, as a developing nation has witnessed rapid urbanization primarily in its Tier-1 cities and the impact of this phenomenon, has adversely affected the living quality in these cities. Various initiatives have been taken up to tackle the urban development issues, respective to the Indian context. India, has recently adopted the concept of Smart City to tackle the above issues but as observed in this study, the respective concept adopted by the Government of India lacks concerns towards the natural environment which is an important dimension of sustainable development of a city; though the mission introduced by the government of India do intend to achieve sustainability in its approach with focus upon economic development, achieving equity with inclusiveness and providing infrastructure services to make urban cities more compact and livable, with the latest addition of the ICT as a tool.

To sum up, it would be prudent to conclude that the School of Construction MBA curricula is ensured significant permeation of sustainability thinking for educating the potential built environment professionals. This thinking is in accordance with the established though of sustainability being allowed to permeate into construction curricula as an integral component, like conventional parameters of cost, time and quality (Conte, 2013)

ACKNOWLEDGMENT

This research received no specific grant from any funding agency in the public, commercial, or not-for-profit sectors.

REFERENCES

Alshuwaikhat Habib, M. A. Y. (2016). Sustainability Assessment of Higher Education Instituions in Saudi Arabia. *MDPI, 8*.

Conte, E. (2013). Sustainability and built environment: The role of higher education in architecture and building engineering. *European Journal of Sustainable Development*, *5*(3), 1–10.

UN Environment and International Energy Agency. (2017). *Towards a zero-emission, efficient, and resilient buildings and.* UNEP.

Raufflet, E. (2013). Integrating sustainability in management education. *Humanities (Washington)*, *2*(4), 439–448. doi:10.3390/h2040439

RICS SBE. (2014). *RICS SBE.* Retrieved August 20, 2018, from http://ricssbe.org/about-us/about-rics-sbe/overview/

Singh, S. R. S. (2018). *Sustainability in Construction Management Education: An Indian Perspective.* Cham: Springer. doi:10.1007/978-3-319-73293-0_9

Tinker, A. A. (2003, April). "Greening" the construction curriculum. *ASC Proceedings of the 39th Annual Conference,* 113-118.

Vanderbilt University. (2018). *Centre for Teaching*. Retrieved from CFT.vanderbilt.edu: https://cft.vanderbilt.edu/2010/08/new-cft-guide-on-sustainability-and-pedagogy/

World Commission on Environment and Development. (1987). *Our Common Future.* UN.

Chapter 8
Integrated BIM Education in Construction Project Management Program

Ki Pyung Kim
https://orcid.org/0000-0003-3018-5504
University of South Australia, Australia

Sherif Mostafa
https://orcid.org/0000-0002-5708-3770
Griffith University, Australia

Kenneth Sungho Park
Aston University, UK

ABSTRACT

The construction industry is currently struggling due to the ever-increasing complexity of a building and the highly fragmented nature of the construction industry. In response to the current problems, building information modelling (BIM) has been adopted to improve productivity and deliver a quality building to a client. The academia also has been striving to embed BIM education into the exiting curriculum to accommodate the needs of the construction industry. However, BIM is currently taught as a simple design tool rather than a methodology to improve productivity in construction projects. Although there have been various attempts to integrate BIM courses into the existing curriculum, a lack of research has been conducted regarding how BIM adoption and implementation strategy should be utilized for a BIM integrated curriculum. Thus, this chapter will provide insights for updating and developing BIM courses.

DOI: 10.4018/978-1-5225-8452-0.ch008

INTRODUCTION

Customer requirements for a building have become more complex and bespoke in its design and materials as well as sustainable attributes related to operational energy consumptions. In order to cater for these requirements and complete a project successfully, effective integration of various construction information such as design, engineering drawing, and energy analysis, and coordination of project management information such as project management plan, quality assurance, and cost and time management among key project participants are more essential than ever. To determine the most affordable building design financially and environmentally, diverse information such as design, costs and environmental impacts of construction materials should be collected and integrated from the outset of a project among key project stakeholders. However, the current practice in the construction industry seems to be challenging to effectively and efficiently cope with the ever-increasing complexity of a project due to the highly fragmented nature of the construction industry and a lack of knowledge and skills of construction personnel to manage stakeholders' requirements.

In response to the current problems, Building Information Modelling (BIM) has been recognized as an enabler that can improve the productivity of a construction project by facilitating collaboration among stakeholders. BIM is defined as an information management system to integrate and manage various construction information throughout the entire construction project life cycle based on a 3D parametric design to facilitate effective communication among project stakeholders to achieve a project goal(s) in a collaborative manner (Kim, 2014). BIM is currently regarded as a major paradigm shift in the construction industry as it is a catalyst for changes of process and culture that requires more integrated approach than before (Hannele et al., 2012). Consequently, various researchers increasingly study a way of utilizing BIM to improve productivity in the construction industry, and the governments mandate the BIM use for the public construction project in the US, UK, Singapore, and South Korea, which is called 'government push and industry pull strategy'. As of 2015, the BIM adoption status indicates approximately 60% in North America, Europe, and Oceania (Jung & Lee, 2015).

The current construction industry is more leaning towards the digitalised Architecture, Engineering, and Construction (AEC) practice due to the BIM capabilities named 'nBIM capability' as a project facilitator that attract the construction industry to adopt BIM in various construction projects over a life cycle effectively and efficiently. nBIM capability can be explained as 3D BIM for visualization of design intent, 4D BIM for project scheduling based on 3D visualization, 5D BIM for project cost management, and 6D BIM for facility management during maintenance and operation phase of a building. Thus, it is evident that the construction industry needs to adopt BIM for improving productivity and gain a competitive advantage on others in the market, and it is essential to retain skilled and experienced BIM personnel to manage a BIM-enabled construction project. Consequently, BIM-ready graduates have become highly desirable as the construction industry wants to minimise the expenses for basic BIM training (BIM in Practice, 2012). Indeed, BIM competence is currently expected for those who want to work in the construction industry (BIM2050, 2014), and various researchers and organisations have identified that proper BIM education will improve the quality of construction projects and enhance the construction industry forward (Suwal & Singh, 2018). Yet, the construction industry is currently suffering from a shortage of skilled construction professionals, and it is much challenging to secure construction professionals who have a proper level of understanding and skills of BIM-based on the knowledge of construction project management.

To tackle the current skill shortage problem and provide relevant knowledge and insights, academia has been striving to restructure and redevelop its curriculum to integrate BIM education into the existing courses and programs. Suwal and Singh (2018) asserted that BIM education should be provided in a multidisciplinary environment where students can learn how to work with others from other disciplines in a collaborative manner. In order to convey proper BIM education, there is a consensus among construction professionals and academics that a series of BIM oriented courses need to be embedded into the overall curriculum rather than a separate stand-alone course, so students can develop relevant BIM skills and knowledge based on construction project management knowledge and practice (NATSPEC, 2017). It has been recognised essential that the construction project management program should provide theoretical knowledge and practical skills to the students, so they can become industry-ready and employable graduates. Based on the studies and research, there have been various initiatives to develop BIM courses to integrate BIM education into the existing curriculum. However, they are mainly an ad-hoc design-oriented education or training that educate students how to develop a 3D building design using a BIM tool such as Revit, ArchiCAD, and Navisworks rather than envisage a big picture how BIM can be used as a construction project methodology for construction professionals to achieve productivity in construction projects from the outset of a project to the operation stage of a building. Moreover, a lack of study has been conducted what kind of BIM adoption and implementation strategy should be utilised for a fully integrated and updated BIM-embedded curriculum in the construction project management program.

Thus, this chapter will provide insights and a practical guide for developing BIM courses and integrating them into the existing curriculum via a systematic review of research on BIM curriculum design, and also a review of actual BIM implemented programs in construction project management programs in various countries. Therefore, this chapter will contribute to identifying and discussing in detail to establish a clear understanding of the status of BIM education nationally and globally. Furthermore, it is expected to provide practical details of BIM curriculum development and its contents in conjunction with administrative and IT supports will be revealed.

BIM AND CONSTRUCTION INDUSTRY

According to a study, productivity in the construction industry has not been improved over two decades while the non-construction industries such as the manufacturing industry have been continuously improved (Teicholz, 2015). It is because the nature of a construction project requires diverse stakeholders over a relatively long time period via various communication mediums, and eventually, the complex project environment makes timely collaboration among key project participants challenging. In response to the current issues in the construction industry, the construction industry is actively adopting and exploiting advanced technologies such as BIM to minimise destructive changes occurred in a later stage of project by visualization of design and construction, and also to improve business efficiencies via cloud-based solutions for timely collaboration (France et al., 2010; Ness et al., 2019). Researchers asserted that the advancements in information digital technologies such as BIM can serve as an enabler to facilitate instrumental changes and improvements in the way of designing, constructing, operating and managing facilities (Kim & Park, 2018; Park & Kim, 2014). Indeed, BIM is being increasingly adopted in the construction industry due to three major benefits: a) Design and Construction Quality Improvement, b) Productivity Improvement (Effective and Efficient Project Information Management) and c) Sustainability Enhancement.

1. **Design and Construction Quality Improvement:** BIM enables design professionals to develop multiple design alternatives, and study these design options regarding cost, time and sustainability. Since BIM use a single data source built into a 3D model based on parametric design, architects, engineers, constructors and other project stakeholders are able to exchange instant feedback on designs and construction methods. This capability can achieve continuous improvement on designs throughout a project life cycle. As all information can be stored and retrieved based on single-source data, improved consistency and integrity in building designs can be achieved, and design conflicts and clash can be detected before the construction phase. Thus, building designs can be developed and tested virtually before the constructors actually build on site, and better quality of designs can be achieved (Froese, 2010).

2. **Productivity Improvement (Effective and Efficient Project Information Management)**: There is considerably less scope for misinterpretation and consequential changes, disruption and reworks (Langdon, 2012; London et al, 2008) as BIM provides a holistic view of building designs, costs, structures and sustainability issues simultaneously based on a single data source. Construction professionals, in particular, facility managers and maintenance teams, can use BIM to manage a building throughout its operational life by providing and retrieving necessary information about maintenance and repair for future use (NHBC, 2013; Eastman, et al., 2011). In addition, researchers (Bryde et al., 2012) studied 20 construction project cases by specifically focusing on the benefits of BIM form the viewpoint of project management. The research revealed that BIM can provide various benefits: cost and time reduction or control, communication and coordination improvement, and quality increase/ control. Thus, a building can be built twice, virtually and actually, which will reduce unnecessary reworks and improve productivity in the construction industry (NHBC, 2013; CIC, 2013a). As a result, BIM can achieve cost reductions of between 15% and 20% (NHBC, 2013) because it can reduce reworks and unnecessary costs by integrating various pieces of information regarding sustainable attributes, and through updating the information as the design changes effectively and efficiently.
3. **Sustainability Enhancement**: BIM supports informed decisions regarding sustainability issues such as energy performance and embodied CO2 of a building at the early design phase by assessing the energy performance and the embodied CO2 of a building where most of the level of sustainability is determined (Redmond et al., 2012; Krygiel and Nies, 2008). However, in a traditional way, it is challenging to secure sustainable attributes of a building from the early design phase because essential information regarding the construction materials that determine embodied CO2 is not readily available. In addition, it is challenging to exchange and integrate essential information amongst project stakeholders, in particular between design and construction professionals due to the fragmented nature of the construction industry. In addition, BIM can achieve sustainability in the construction supply chain by making construction and procurement processes more effective and efficient, and construction waste materials throughout a project life cycle can be reduced (HM Government, 2012; Crosbie et al., 2011)

Recently, the construction industry is utilizing BIM more broadly based on convergence of Industry 4.0 technologies and BIM such as Big Data analytics with BIM for better design alternatives, AI-based architectural design and construction with BIM for better quality construction, and VR/AR application with BIM for safer working environment for construction professionals (Kim et al., 2018) As a result, the governments in various countries mandate the BIM use for the public construction project in the US, UK, and South Korea since 2006 and 2016 respectively, and BIM-based construction project management has become daily processes and tasks as a de facto standard for public and large projects where large investments are made (Kim et al., 2016).

GLOBAL TREND OF BIM EDUCATION AND CURRICULUM

In response to the government mandates and the demands from the construction industry, there has been a new emerging role in the construction industry such as BIM manager, BIM coordinator, and BIM modeller as mega projects such as Capital Gate and Burj Khalifa in the United Arab Emirates around the world require the BIM-enabled personnel who can manage a BIM-based construction project. The evolution of BIM initiates the changes of traditional roles of construction professionals, and the changes currently affect education and training of construction project management methodologies (PAS, 2016). Researchers (Bew & Underwood, 2010) argue that digital transformation from the traditional way of management is difficult as digital information management processes that can facilitate the digital transformation are not readily available and there are not enough personnel to educate new digital technologies in conjunction with proper construction project management processes. Various research on BIM education have been conducted (Abdirad & Dossick, 2016), and researchers identified that a lack of proper BIM education and training are still hampering down the construction industry to fully adopt and utilise BIM for construction projects (Yalcinkaya & Singh, 2015). Thus, it is evident and instrumental that BIM education should be taught in conjunction with other courses such as project management or construction management rather than a stand-alone course to envisage the workflows in a BIM environment and to foster students' ability to manage a BIM-based construction project properly. Indeed, the International Organization for Standard (ISO) released the ISO 19650-1 and ISO 19650-2 for the information modelling method in a BIM environment, and this standard explicitly indicates that information modelling method standard is included in the international standard for project management (ISO, 2012) as one component to deliver a required project outcome to a client. Recent research identified that basic skills and knowledge associated BIM are highly required in the current construction industry job market for both undergraduate and graduate levels (McLernon et al., 2015).

Furthermore, BIM maturity of the construction industry needs to be enhanced in order to adopt and utilise BIM without creating more confusion and problems in the way to integrate BIM for the existing practices among diverse project participants (Succar, 2010). Tanyer (2010) argue that an integrated BIM education and training is essential for successful BIM implementation to improve the productivity of construction projects and accommodate the BIM-enabled environment better among key project participants. Hence, it is an essential pre-requisite to secure construction professionals equipped with BIM knowledge technologically and methodologically in order to improve BIM maturity within an organization. However, the construction industry is suffering from a lack of experienced BIM-ready personnel, and it has become challenging to improve maturity. Consequently, a vicious cycle between a lack of maturity and a lack of adequate human resource has been presented, and only proper training and education can break the cycle by providing BIM-ready graduates to the construction industry.

National Strategies for BIM Education

As aforementioned, the BIM education should change culture and work processes in the construction industry as well as update a technical skill enabling construction professionals to utilise a BIM tool. Furthermore, diverse project participants get involved in a construction project cross disciplines and cultures, and the diversity demands BIM to serve as a new project delivery process and methodology based on the same terminologies and workflows stemmed out from the same strategy. In order to organize and manage a BIM-embedded project, nation-wide BIM education strategies have been developed to educate students and industry professionals based on the same knowledge for BIM as shown in Table 1.

As of 2013, the UK has 30 higher education institutions and training centres delivering BIM education either an integrated BIM course with other courses or a stand-alone BIM course (HEA, 2013; Heesom & Boden, 2016). Most of the institutions provide the BIM courses under a Master degree program, and recently, the University of Wolverhampton in the UK commenced educating BIM in Undergraduate degree program in the Architectural school. In alignment with the UK national BIM education strategy, professional construction bodies such as Royal Institute of British Architects (RIBA), Royal Institution of Chartered Surveyors (RICS), and Institution of Civil Engineers (ICE) are also providing BIM training for the construction industry. RICS (2015) published 'BIM for cost managers: requirements from the BIM model' to provide professional guidance for a quantity surveyor or a cost manager regarding a building information model development and quality check of cost information. In 2017, 'Building Information Modelling for Project Managers' has been released to support a project manager

Table 1. National BIM education strategy

Country	BIM Education Strategy	Remark
Australia	'National BIM initiative', buildingSMART Australasia, 2012	A BIM education taskforce based core curriculum delivery identified in the support of the National Building Specification (NATSPEC)
	'BIM Education, BIM in Practice', Australian Institute of Architects, 2012	'Users of BIM Education', 'BIM Education Providers', and 'Draft Collaborative Education Framework' are defined
	'BIM knowledge and Skills Framework', Australian Construction Industry Forum and Australasian Procurement and Construction Council, 2017	A framework with a proper guide and a base to establish a BIM curriculum based on the same principles, practices and terminologies, and to assess students' learning outcomes and progress.
UK	'National qualifications frameworks (updated)', BIM Academic Forum (BAF), 2013	Learning outcomes of BIM education defined at undergraduate (Level 4, 5, and 6) and postgraduate programs (Level 7) in accordance with national qualification frameworks. BIM teaching impact matrix identified with BIM level: Absent, Aware, Infused and Embedded. The framework is developed based on a long-term vision to foster BIM skills and knowledge.
Canada	'Roadmap to Lifecycle BIM', buildingSMART Canada, 2014	Lifecycle BIM roadmap for a collaborative BIM education with a two-step approach: a) Developing a curriculum for BIM education and accreditation, b) Developing BIM training packages providing certification
	'Industry-wide benchmarking platform', Canada BIM Council, 2016	four levels of Certification for Individuals as well as Educational Course and Program Certification

to understand how BIM can be integrated with project management methodology and processes (RICS, 2017). In Australia, there have been various developments for BIM education strategy including a) 'Program-wide Implementation Strategy' for BIM education in Tertiary Construction Project Management program at University of Technology Sydney (Foresythe et al., 2013); b) 'Global Passport through Co-integration of Construction Immersive Environments' at RMIT University (London, 2015); c) 'Collaborative Design Approaches to Teaching BIM (CODEBIM)' based on a collaboration among three institutions - University of South Australia, University of Newcastle and University of Technology Sydney. In Canada, there was a recent workshop for BIM education with a collaboration of five institutions from four different countries - University of Liverpool (UK); University of Technology Sydney (Australia); Drury University (US); Ryerson University and University of British Colombia (Canada). The workshop was concluded with six key actions for BIM education that can bridge a gap between a theory and a practice of BIM (Poirier et al., 2016).

BARRIERS FOR INTEGRATED BIM EDUCATION

Based on the identified BIM education strategies at the previous section, it is evident that the various BIM strategies have two major aspects in common – a) BIM course integration into the existing curriculum program-wide; b) Focuses on the collaborative workflow of a project. Although a properly integrated education for BIM is essential, there has limited research conducted how to provide an integrated BIM education for students, and how to assess students' performance and the level of understanding about the BIM-based construction project management. Moreover, BIM education and training are currently provided by professional training companies, academic institutions or professional construction bodies falling into the following major categories – a) Basic ad-hoc BIM Tool Training; b) Advanced ad-hoc BIM Tool Training; c) Fundamental BIM Knowledge focusing on Theory and Case Studies; d) BIM standards training such as PAS 1192 series or ISO 19650 series. There is a lack of proper integrated BIM education or training that can glue a technical aspect onto a theoretical/managerial aspect of BIM (Kocaturk & Kiviniemi, 2013; Abdirad & Dossick, 2016). Currently, the construction industry is seeking for practical applications of BIM that can render immediate benefits, and it is challenging to fully adopt BIM technically and methodologically (Forgues & Iordanova, 2010). Indeed, the NATSPEC emphasised that the important role of BIM education is not only to educate technical skills using BIM software, but also to enlighten students and construction professionals on the collaborative workflow within a BIM-embedded project environment as well as technical skills (Rooney, 2015; Hsieh et al., 2015). Since BIM is not a purely theoretical subject, it must have practicality based on a managerial framework and work processes. This is confirmed by researchers (Abdirad & Dossick, 2016; Puolitaival & Forsythe, 2016) that the construction industry expects technical and non-technical aspects of BIM, and currently, non-technical skills have become increasingly important. In order to educate technical and managerial BIM synchronously, lessons learned from real-life projects should be utilised to bring students a sense how a real-life BIM model project looks like and how stakeholders can collaborate throughout a project lifecycle (Lostuvali et al., 2010; Underwood & Isikdag, 2010; Riese, 2010). In this way, the hands-on experience on a real-life BIM project will contribute for students and construction professionals to develop a practical sense of BIM (Gier, 2015; Peterson et al., 2011). Despite the importance and relevance of BIM education based on industry-academia collaboration, it is challenging to obtain a real-life BIM project information and a building model due to the Intellectual Property issues. Alternatively, a private design company can be used to procure a BIM model, but it costs too much and the model present too complicated design or too much information for those who just start learning BIM (Clevenger et al., 2010). Although a BIM model can be simplified in design with less information, there is not a proper person to specifically demand the requirements and actually check the BIM model when it is delivered. The issue has been recognised by Cheng (2006), and the researcher emphasised the level of expertise is crucial to deliver proper BIM education. Becerik-Gerber et al. (2011) pointed out a lack of qualified personnel to educate BIM and a lack of inadequate BIM accreditation standard for BIM curriculum (Puolitaival & Forsythe, 2016).

Another barrier to BIM education is a lack of support as known as internal buy-in from others. Currently, BIM software can be obtained for free from a software vendor with an academic license. However, hardware that is capable of operating BIM software properly requires an upfront initial investment (Hogle, 2013; Magiera, 2013). Macdonald (2012) pointed out the importance of internal buy-in and support from a management group because the roles and responsibilities must be clearly defined and assigned in order to make an initial investment, in other words, it must be clearly arranged who will take the financial investments. Furthermore, once BIM equipment is set up, a BIM course or BIM curriculum needs to be developed. However, it requires a working group to integrate BIM courses into the existing curriculum, and develop a new BIM integrated curriculum. It seems a small change as BIM courses will be added to the existing curriculum, however, academics need to discuss how to embrace BIM aspect into the existing courses and assessments need to be modified if necessary. In addition, administrative staffs also need to reflect the changes into an online program management system (Becerik-Gerber et al., 2011; Rooney, 2017). Thus, it is not possible to establish an integrated BIM education without internal buy-in from other colleagues and clear roles and responsibilities among academics and administrative staffs (Suwal & Singh, 2018). This chapter will contribute to providing practical details of BIM curriculum development and its contents in conjunction with administrative and IT supports will be revealed.

RESEARCH METHOD

Research strategy guides researchers to plan a process to conduct researches to answer the research questions (Saunders et. al, 2009; Bryman, 2008). Yin (2003) asserts that a researcher needs to consider three major aspects to select relevant research strategy - Form of research questions, Necessity of control of behavioural events, Focus on contemporary events (Yin, 2003) as summarised in Table 2.

This research requires no control over behavioural events as a researcher cannot intervene academics and administrative staffs' attitudes toward the BIM education development, and focuses on a contemporary event which is an integrated BIM education development to meet the BIM-ready graduate demands from the construction industry and improve productivity in the construction industry. Interactions among

Table 2. Relevant situations for different research strategies (Yin, 2003)

Strategy	Form of Research Question	Requires Control of Behavioural Events?	Focuses on Contemporary Events?
Experiment	How, why?	Yes	Yes
Survey	Who, what, where, how many, how much?	No	Yes
Archival analysis	Who, what, where, how many, how much?	No	Yes/No
Case study	How, why?	No	Yes

academics, students and administrative staffs and rationale behind action plans for BIM education development can be identified by asking how and why. Thus, this research adopted a case study as the most relevant research strategy.

The case study focuses on the first and fourth (final) year students in the Construction Project Management program at the School of Natural and Built Environments (NBE), University of South Australia (UniSA). The first and fourth-year students are chosen as the first and fourth years are the time when students have maximum exposure to a BIM environment, and the most of fourth-year students are employed in the construction industry. As shown in Table 3, the detailed information about the BIM courses for the first and fourth years are provided.

Specific students are selected from the courses based on the following criteria – a) Complete the BUIL1021 Construction Communication course in the 1st year, and achieved overall grade above Credit with Distinction on a BIM assessment; b) Enrolled in the BUIL4027 Integrated Project with minimum 100day working experience in the construction industry, and employed at present working on a construction project. The specific requirements for recruiting students are essential as the relevance and importance of BIM education can be confirmed by those students who have been studied technical and managerial BIM from the beginning

Table 3. BIM courses in the 1st and 4th year courses

Course Number	Course Name	Course Level	Description
BUIL1021	Construction Communication	1st year, Undergraduate	Basic BIM Knowledge and BIM Software Skill are educated in conjunction with the Construction Project Management Processes.
BUIL4027	Integrated Project	4th year, Undergraduate	A Capstone Project Course for students to practice integrating Construction Project Management Processes with BIM knowledge and skills using a real-life BIM model.

to the end of the academic lifecycle. In addition, the currently employed students can provide more relevant insights and comments on the BIM education established in the Construction Project Management program. Consequently, 31 students were selected to identify how they think about the BIM courses, and how BIM courses are useful for them to do their tasks. In order to identify students' thoughts, a focus group interview has been conducted.

BIM EDUCATION IMPLEMENTATION STRATEGY

The School of NBE promoted the importance of BIM education, and establishes an internal buy-in for BIM course integration from academic and administrative staff members by the following steps.

Step 1: Staffing for BIM course development: There was an existing course, BUIL1021 Construction Communication, designed to educate students about the technical and managerial aspects of BIM. However, there was no BIM expert who has the proper theoretical and technical knowledge and experience of BIM. As aforementioned, it is instrumental to secure a BIM expert who has the ability to teach and develop a BIM course with proper knowledge and skills. As a result, a BIM expert was hired who has proven his technical skills and managerial knowledge of BIM through winning an award at a global BIM competition hosted by buildingSMART, and also his PhD study in BIM. Furthermore, the BIM expert completed his Master degree in Project Management and possessed multiple Project Management professional credentials, which is another important point to be considered for proper BIM education development as aforementioned in the Barriers for Integrated BIM Education section.

Step 2: Forming a BIM Working Group: Once a BIM expert was hired, the School of NBE defined roles and responsibilities for BIM course development and integration. A BIM Champion who support decision and procure resources for BIM course development and integration has been assigned to a senior member such as a Program Director and a Discipline Leader, and a BIM expert, who is a lecturer as a subject matter expert, took the responsibility to develop a new BIM teaching materials and modify the existing course structure of the BUIL1021 Construction Communication to accommodate relevant BIM aspects into the course. Furthermore, the existing assessment has been modified and updated to assess students' understanding of managerial BIM aspects and also technical skills of BIM software such as Autodesk Revit.

Step 3: Regular Program Meetings: The updates and modifications in the BIM assessment must be reflective on the online program management systems to update a Course Outline and a Course Assessment Result Entry. Subsequently, the BIM working group collaborates with the administrative staff members to update the information in the systems and arrange a regular program meeting with academics who may or may not be affected by the BIM course development and integration. Through the regular meetings, all academics and administrative staffs are fully informed about the progress on the development, and the meetings served as an opportunity to discuss any concerns or suggestions.

Step 4: Specific Course Alignment Meetings: Apart from the regular meeting, BIM working group and course coordinators for specific courses that relate to BIM course had a meeting to align the BIM courses with other courses.

Step 5: IT Support and BIM Training Support: BIM software such as Autodesk Revit, Autodesk Navisworks, and CostX were installed by an IT department, and the PCs have been upgraded specifically for the BIM software. As aforementioned, initial upfront investment on PCs and software is one of the important barriers to achieve a proper BIM education development. However, commitment from the management for BIM education secures the IT support. In addition, an up-to-date BIM training for a BIM expert and other academics were provided to bring the understanding of BIM of academics up to the level where they can effectively integrate relevant course specific BIM aspects into their courses such as Quantity Surveying, Construction Management Fundamental, Construction Scheduling, and Capstone Project.

BIM Course Integration Result

As shown in Table 4, the BIM aspects that have been taught in the program were maintained with more up-to-date BIM information, and the courses that did not include any BIM aspects, although there is a positive potential to include BIM aspects, were developed further and updated the course contents and assessments if relevant.

However, there is a certain course difficult to implement BIM aspects such as Building Surveying, Contract Administration and Construction Law. The current Building Surveying practice has not embraced BIM to confirm whether a building design complies with the building codes. The current practice is mainly based on a paper drawing based, and it makes difficult to teach BIM for Building Surveying. Furthermore, BIM cannot be a panacea for all the construction project management subjects as there is a specific course such as Construction Law and Contract that cannot be fathomed by a simple BIM software. Although there is the Integrated

Project Delivery as a project delivery method for a BIM construction project, there are many other topics that are not related to BIM. Thus, certain courses are excluded from the BIM course integration.

Table 4. Result of BIM course integration into the program

Course Name	Before BIM Implementation	After BIM Implementation
First Year		
Construction 1	No	BIM Guest Lecture from Industry Partners
Introduction to Construction Management	Basic BIM Introduction	Basic BIM Introduction
Construction Communication	Basic BIM Introduction	Basic and Advanced BIM Introduction, BIM Software (Autodesk Revit) Education, Applied ICT in conjunction with BIM such as AR/VR/3D Scanner Introduction and Demonstration
Second Year		
Quantity Surveying Practice 1	CostX for Quantity Surveying without BIM aspects such as a BIM model manipulation	CostX with Basic BIM Introduction for Quantity Surveying
2nd Semester		
Construction Scheduling	No	BIM for Time Management (4D BIM Introduction)
Building Services N	Clash Detection using BIM software (Autodesk Navisworks)	Clash Detection using BIM software (Autodesk Navisworks)
Third Year		
Construction Environmental Science	No	Basic Demonstration of BIM software (Autodesk Revit)
Sustainable Construction	No	Advanced BIM Software for Energy and Sustainable Design (IES Virtual Environment)
Fourth Year		
Integrated Project	BIM Software for Design Purpose	Capstone Project for a BIM-enabled Construction Project using a real-life BIM Model. Advanced BIM Education as a technical tool and a managerial process. Design Alternative Analysis using BIM software (Autodesk Revit), Building Code Checking and Risk Management Plan Development using BIM software (Autodesk Navisworks), Cost Plan for Construction using BIM software (CostX)

Focus Group Interview Result

Due to the time constraint and work commitments, 31 students were divided into two groups and two focus group interviews were conducted. The two groups discussed the two BIM courses, and two major discussions have been made: a) Overall Satisfaction for the Course Contents, and b) Relevance of the Course Contents and Structure. All students were satisfied with the course contents and they unanimously agreed that they satisfied with the BIM course materials and teaching style. The followings are the students' comments.

Student A: "*This course has helped me to gain a better understanding of BIM.*"
Student B: "*My knowledge of BIM and in particular Revit has grown and I know believe that I could successfully teach someone how to use the software.*"
Student C: "*The learning of BIM will be very beneficial, the lecturer had a good knowledge of this.*"
Student D: "*Working with BIM was very interesting*"

Regarding the relevance of BIM course contents and structure, students indicated that they are happy to learn BIM knowledge and skills, which help them to do their jobs. Also, students appreciated a real-life BIM construction project for a capstone project since they can see how their skills and knowledge have been improved. The followings are the students' comments.

Student A: "*The Hanson Yuncken (Australian Construction Company) guest lecture was extremely relevant and found it to be helpful with the course.*"
Student B: "*BIM is an important part of ICT in the construction industry so learning how to use it is a good skill to have.*"
Student C: "*I like his ways of teaching the students in technologies software; I think he is the best lecturer for technology parts (BIM, Revit, Primavera, and Project scheduling).*"
Student D: "*I liked how he made improvements to the assessments and I felt like I actually got a lot out of this course. I liked doing the feasibility study and Ki asked our group good questions to keep us on our toes.*"

Based on the interview, the BIM course integration and BIM course development were proven successful, and students are satisfied with what they studied and learned.

CONCLUSION

Academia has been striving to embed BIM education into the existing curriculum to accommodate the needs of the construction industry. Although there have been various attempts to integrate BIM courses into the existing curriculum, a lack of research has been conducted regarding how BIM adoption and implementation strategy should be utilised for a BIM integrated curriculum. Based on the case study of the Construction Project Management program in the School of Natural and Built Environments, University of South Australia, this research provides a practical guide for the BIM education implementation strategy through a step-by-step process. Also, this research identified which subject can embrace BIM aspects and which subjects cannot due to the uniqueness of the subject. Finally, this research recognised students' appreciation and satisfaction from the BIM courses, and it has been revealed that students would like to hear about a real-life BIM project from construction professionals. Furthermore, students' satisfaction becomes higher when they put their hands on a real-life BIM construction project using a real-life BIM model. Thus, this chapter will provide insights for integrating new and updated BIM courses into a program, and also enlighten professionals and academics who seek for a practical tip to successfully develop a BIM integrated program based on a collaborative manner between academics and administrative staffs.

REFERENCES

Abdirad, H., & Dossick, C. S. (2016). BIM curriculum design in architecture, engineering, and construction education: A systematic review. *ITcon*, *21*, 250–271.

Abdirad, H., & Dossick, C. S. (2016). BIM curriculum design in architecture, engineering, and construction education: A systematic review. *Journal of Information Technology in Construction*, *21*(17), 250–271.

Barlish, K., & Sullivan, K. (2012). How to measure the benefits of BIM—A case study approach. *Automation in Construction*, *24*, 149–159. doi:10.1016/j.autcon.2012.02.008

Becerik-Gerber, B., Gerber, D. J., & Ku, K. (2011). The pace of technological innovation in architecture, engineering, and construction education: Integrating recent trends into the curricula. *ITcon*, *16*, 411–432.

Becerik-Gerber, B., Gerber, D. J., & Ku, K. (2011). The pace of technological innovation in architecture, engineering, and construction education: Integrating recent trends into the curricula. *Journal of Information Technology in Construction*, *16*, 411–432.

Bew, M., & Underwood, J. (n.d.). *Delivering BIM to the UK market*. doi:10.4018/978-1-60566-928-1.ch003

BIM2050 Group. (2014). *Built Environment 2050 – A Report on Our Digital Future*. London: Construction Industry Council.

Cheng, R. (2006). Suggestions for an Integrative Education. In N. Strong, D. Friedman, & M. Broshar (Eds.), *Report on Integrated Practice*. Washington, DC: AIA.

Forgues, D., & Iordanova, I. (2010). An IDP-BIM Framework for Reshaping Professional Design Practices. In *Construction Research Congress 2010* (pp. 172-182). American Society of Civil Engineers. 10.1061/41109(373)18

Handbook of Research on Building Information Modelling and Construction Informatics. (2010). *Concepts and Technologies*. Hershey, PA: IGI Global.

Hannele, K., Reijo, M., Tarja, M., Sami, P., Jenni, K., & Teija, R. (2012). Expanding uses of building information modeling in life-cycle construction projects. *Work (Reading, Mass.)*, *41*, 114–119. PMID:22316709

Hogle, L. (2013). *Industry + Academia: The perfect partnership*. Paper presented at the BIM Academic Education Symposium. Setting the Course for a BIM Educational Strategy, Washington, DC.

Jaques, P. A., & Viccari, R. M. (2006). Considering students' emotions in computer-mediated learning environments. In Z. Ma (Ed.), *Web-based intelligent e-learning systems: Technologies and applications* (pp. 122–138). Hershey, PA: Information Science Publishing. doi:10.4018/978-1-59140-729-4.ch006

Kim, K. P. (2015). *Conceptual Building Information Modelling Framework for Whole-House Refurbishment Based on LCC and LCA* (Unpublished doctoral dissertation). Aston University, Birmingham, UK.

Kim, K.P., Ma, T., Baryah, A.S., Zhang, C., & Hui, K.M. (2016). Investigation of readiness for 4D and 5D BIM adoption in the Australian construction industry. *Management Review: An International Journal, 11*(2), 43-64.

Kim, K.P., Ochoa, J.J., & Kwon, D. (2018). Feasibility study of BIM integration with big data for construction scheduling. *Journal of Advanced Research in Dynamical and Control Systems, S11*, 1260-1265.

Kim, K. P., & Park, K. S. (2018). Delivering value for money with BIM-embedded housing refurbishment. Facilities, 1-18. doi:10.1108/F-05-2017-0048

Kocaturk, T., & Kiviniemi, A. (2013). Challenges of Integrating BIM in Architectural Education. *Proceedings of the 31st International Conference on Education and research in Computer Aided Architectural Design in Europe.*

Kocaturk, T., & Kiviniemi, A. (2013). Challenges of integrating BIM in architectural education. Paper presented at the eCAADe 2013: Computation and Performance, Delft, The Netherlands.

Kugbeadjor, W., Suresh, S., & Renukappa, S. (2015). BIM awareness and readiness of postgraduate built environment students in West Midlands Universities, UK. In *CIB International Conference Proceedings. Going North for Sustainability: Leveraging Knowledge and Innovation for Sustainable Construction and Development* (pp. 531-543). London: IBEA Publications.

Lee, N., & Hollar, D. A. (2013). Probing BIM education in construction engineering and management programs using industry perceptions. In *49th ASC Annual Int. Conf. Proc.* California Polytechnic State Univ.

Leite, F., & Wang, L. (2014). Process-Oriented Approach of Teaching Building Information Modeling in Construction Management. *Journal of Professional Issues in Engineering Education and Practice, 140*(4), 4014004. doi:10.1061/(ASCE)EI.1943-5541.0000203

Macdonald, J. A. (2012). *A framework for collaborative BIM education across the AEC disciplines.* Paper presented at the 37th Annual Conference of Australasian University Building Educators Association (AUBEA).

Magiera, J. (2013). Integrated teaching for integrated engineering practice-BIM on campus. Paper presented at 210th International Conference on Computer Methods in Mechanics, Poznan, Poland.

Miller, G., Sharma, S., Donald, C., & Amor, R. (2013). Developing a building information modelling educational framework for the tertiary sector in New Zealand. In *IFIP International Conference on Product Lifecycle Management* (pp. 606-618). Springer. 10.1007/978-3-642-41501-2_60

Ness, D., Kim, K., Swift, J., Jenkins, A., Xing, K., & Roach, N. (2019). Cradle to cradle building components via the cloud: a case study. In *6th CIB International Conference on Smart and Sustainable Built Environments - SASBE2018*. Emerald Publishing Group.

Park, K. S., & Kim, K. P. (2014). Essential BIM input data study for housing refurbishment: Homeowners' preferences in the UK. *Buildings, 4*(3), 467–487. doi:10.3390/buildings4030467

Productivity, C. (2015). *Teicholz*. Retrieved from https://projectproduction.org/presentations/2015/2015-PPI-Teicholz.pdf

Puolitaival, T., & Forsythe, P. (2016). Practical challenges of BIM education. *Structural Survey*, *34*(4/5), 351–366. doi:10.1108/SS-12-2015-0053

Rooney, K. (2017). *BIM education - global. 2017 update report*. ICIS Integrated Construction. Retrieved May 28, 2017, from http://www.canbim.com/single-post/2017/05/29/NATSPEC-Global-BIM-Education-Report-2017

Rooney, K. (2018). *NATSPEC Construction Information; BIM Education Global 2018 Update Report*. Retrieve from https://www.icis.org/wp-content/uploads/2018/05/BIM-Education-Global-2018-Update-Report-V5.0.pdf

Sabongi, F. J. (2009). *The integration of BIM in the undergraduate curriculum: An analysis of undergraduate courses*. Paper presented at the International Proceedings of the 45th ASC (Associated Schools of Construction) Annual Conference, Gainesville, FL.

Sacks, R., & Pikas, E. (2013). Building information modeling education for construction engineering and management. I: Industry requirements, state of the art, and gap analysis. *Journal of Construction Engineering and Management*, *139*(11), 04013016. doi:10.1061/(ASCE)CO.1943-7862.0000759

Sawyer, S., & Tapia, A. (2005). The sociotechnical nature of mobile computing work: Evidence from a study of policing in the United States. *International Journal of Technology and Human Interaction*, *1*(3), 1–14. doi:10.4018/jthi.2005070101

Suwal, S., & Singh, V. (2018). Assessing students' sentiments towards the use of a Building Information Modelling (BIM) learning platform in a construction project management course. *European Journal of Engineering Education*, *43*(4), 492–506. doi:10.1080/03043797.2017.1287667

Teicholz. (2015). *Construction Productivity Index Graph*. Indexes of labor productivity from construction and non-farm industries, 1964 -2004, indicates no productivity gain in the construction industry.

Yalcinkaya, M., & Singh, V. (2015). Patterns and trends in building information modeling (BIM) research: A latent semantic analysis. *Automation in Construction*, *59*, 68–80. doi:10.1016/j.autcon.2015.07.012

Chapter 9
OnSite:
The Virtual Site Visit as an Environment for Construction Learning

Robert Sean Pickersgill
University of South Australia, Australia

Rameez Rameezdeen
University of South Australia, Australia

Jennifer Harvey
University of South Australia, Australia

ABSTRACT

The chapter summarizes the educational pedagogy researched and developed in the OnSite project, a multi-year trial of a blended virtual learning environment, situating it in the context of immersive learning environments generally and discussing the specific challenges in designing and creating environments suitable for introductory construction courses. It documents and reports the challenges in creating a virtual learning environment (VLE) for use within an introductory construction course for architecture and building students at the University of South Australia. In addition, the chapter will reflect on issues of technical development for immersive learning environments, discussing the purpose and value of high-fidelity modelling, texturing, and lighting to achieve learning "authenticity." Finally, the chapter looks at the implications for VLEs of this sort in terms of larger issues regarding the potential for game engine (GE) environments as cooperative spaces within the AEC industry.

DOI: 10.4018/978-1-5225-8452-0.ch009

INTRODUCTION

A total of 280 students over a three-year period were involved in the *OnSite* Project within the School of Art Architecture and Design at the University of South Australia, undertaking it as a component of their first-year introductory course on Australian domestic construction practices. Courses of this nature tend to be similar across Australian architecture and building courses, and *OnSite* was designed to be applicable, even with regional differences, across the country. A number of the questions related to Australian Standards of building practice that it was important for students to be introduced to.

Within introductory construction courses, the core body of knowledge concentrates on the student acquiring familiarity with conventional domestic construction practices. This information usually assumes that students have no prior knowledge and, as such, need to be introduced to the simplest and most ubiquitous methods being undertaken. While, in Australia, there are some regional differences regarding the use of masonry and timber frame methods, Australian Standards govern all construction methods and levels of performance and are consistent nationally.

METHODOLGY

Two questions emerge from this situation: How does an educator meaningfully address the structural requirements of 'seeing' the reality of domestic construction processes, and; what alternatives exist for this experience if a physical site visit is not feasible? It was the opinion of the authors that developments within Virtual Learning Environments (VLEs) in education would provide a model for simulated experience that successfully managed the manifold challenges of 'real life' learning while also allowing for focussed pedagogical aims that successfully measured the effectiveness of the process. While literature on the subjects of VLEs, gamification, constructivist learning and LMS strategies is currently broad (Zheng, 2015), there is currently little substantive analysis of projects that attempt to syncretically incorporate the full range of capabilities within the digital education space for architecture and construction. Maghool, Moeini and Arefazar, have identified and summarized the relationship between learning styles within architectural education and the potential implementation of VLEs (Maghool et al, 2018). Their LADUVR project addressed similar concerns as those that have premised OnSite, but there are significant differences that OnSite addresses, in particular those relevant to site visits and to the parallel incorporation of LMSs. Projects tend to concentrate upon the employment of gamification methodologies at the expense of visual authenticity, rehearsing a philosophical question that exists within digital game theory of the difference between ludological (instrumental play behavior) versus narrative (immersiveness) approaches. Our challenge was to combine the incorporation of complex software uptake as a condition of learning; the creation of VLEs that demonstrate visual authenticity by employing contemporary game engines; integration between VLEs and learning management systems (LMS); and the application of this material to large data sets (in excess of 30 participants). The work of Maghool, Moeini and Arefazar is clearly a front-runner in the development of a complex and integrated approach to the creation of meaningful and authentic emulation of the construction process, however in our view much of the literature in this field does not attempt to create a fully integrated approach. This was the aim of the *OnSite* project. To progress this aim, a number of questions had to formulated and an appropriate methodology be developed.

The first question that needed to be asked concerned the development of alternative educational experiences: specifically, can virtual learning environments meaningfully substitute for 'real world' experiences? Secondly, if the answer is affirmative for the reasons set out below, how best can the capacities of VLE and the software available be employed to support and reinforce learning objectives? The *OnSite* project addresses the issue of developing an effective method for introducing students to the principles of domestic construction by developing a virtual learning environment (VLE) that employed best practice in combining information-rich content within an immersive and authentic virtual environment. The steps to achieve this required an understanding of the effectiveness of employing a VLE within an LMS; the development of an information-rich VLE that demonstrated authentic learning experiences; and the integration of appropriate assessment modules integrated with the VLE experience. In short, the research group had to develop virtual environments that emulated the experience of construction site visits, employ the most accessible and seamless software suitable for distribution to a broad range of digital platforms (student computers/devices), and meaningfully incorporate and assess their experience of the environments to ensure they concentrated on the pedagogical content.

BACKGROUND

A convention of introducing students to this body of knowledge is the assumption that a site visit, undertaken ideally at regular intervals, will allow students to familiarize themselves with the built reality of how these methods take place. In part this stems from a prior tradition of apprenticeship in which trade craft was learnt at the side of a skilled craftsperson, observing the methods of construction and acquiring knowledge of what constituted best practice. While this form of apprenticeship is still common within the VET system (Vocational Educational Training) and formal apprentice relationships through TAFE (Technical and Further Education), for students studying tertiary degrees in architecture and building it is not feasible for them to devote the amount of time necessary to acquire this form of first-hand knowledge.

For this reason, tertiary degrees adopt a compromise situation in which some familiarity with the trade craft is desirable and is intended as a foundational knowledge of construction principles. Further, for architecture students, there is the expectation within most degrees that this knowledge will form the basis of a skill set that will encourage students to explore other forms of constructional methods. Architecture programs generally privilege the development of creative skills that, while based on convention, encourage innovation where appropriate. However, to achieve this level of innovation there has to be a body of knowledge from which to commence. It is in this context that first-year construction courses aim to introduce this knowledge, tacitly expecting that the academic content of the course will be essentially about acquiring familiarity without expecting innovation at this stage.

Typical also is the parallel pedagogical aim to introduce students to the variety of media within which constructional knowledge is documented and communicated. Within the professional context of designers communicating to tradespeople, the medium of communication, the constructional drawing, retains a series of important conventions. Usually, construction drawings communicate the scale, extent, method, and material reality of a design. While this is a commonplace for any person with a passing familiarity with the industry, most tertiary courses in architecture and building will not assume that this is the case. For this reason, instructing students in the causal relationship between a construction drawing and its built reality is an important phase of the student understanding how construction occurs. Of particular focus, and with reference to the idea of the site visit, understanding and appreciating how a two-dimensional drawing is translated into a three-dimensional reality is a core aim of these courses.

THE *ONSITE* PROJECT

In the tertiary course upon which the *OnSite* project, discussed below, is situated, the course material is part of a suite of learning experiences that also incorporate the acquisition of a number of other relevant skills and learning content. The course contained three assessment points: a vertical sculpture in which students constructed the highest open frame structure from a single sheet of balsa; the design and documentation of a small pavilion, in which students amended and documented a small pavilion using examples of drawing and documentation protocols, and; the *OnSite* virtual learning environments and quizzes. Respectively, these tasks introduced students to the idea of structure (the vertical sculpture), the relationship between

Figure 1. Sample student model from the first year construction course illustrating frame exploration (UniSA, 2017)
Source: UniSA, 2017

two-dimensional orthographic drawings and a small building (the pavilion) (Figure 1), and the three-dimensional reality of how construction takes place (*OnSite*).

All three assessment points were designed to utilize different media of communication. The value of *OnSite* was that it described a relationship between orthographic representations and the three dimensional VLE of the game space. Students were given construction drawings of aspects of the modelled environment and asked to compare, as part of the exploratory process, the difference between the drawings and the 'built' reality, a practice that would have been appropriate if the traditional form of site visit had been available.

On the matter of site visits generally, the course assumed modest or no prior knowledge, the *OnSite* project was intended to introduce students to the scope of domestic construction, test them on knowledge appropriate to an introductory course, and to provide an understanding of the complex, three dimensional nature of construction sites. In previous iterations of the course, students were asked to consistently visit current construction sites and maintain a visual diary of the progress of the works. The intent of this arrangement was to ensure that the element of material reality was made apparent to the student and they recognized the relationship between drawings and the various trades they observed. However, this arrangement had become increasingly difficult by 2014 for a number of reasons:

1. **Consistency**: Without taking on the organizational role of ensuring that all students attended the same site locations at the same time, it was difficult to guarantee consistency of experience.
2. **Coordination**: Because the teaching period typically covered a 13-15 week period, the challenge to find an appropriate site that could demonstrate the informative levels of constructional development was always present. In

addition, there was little guarantee that the constructional schedule of the site could be coordinated with the delivery of the curriculum to the students. At times student's knowledge was in advance of, or behind, what they were viewing on the site visit.

3. **Organizational Complexity**: To ensure that there was a sufficient number of site visit experiences, it would have been necessary for the responsible academic to liaise with the builder/contractor for access to the site at the appropriate times.
4. **OHSW**: the presence of students on a construction site created issues regarding public liability and site safety for the builder/contractor and the university. In addition, when site visits were to be arranged for classes of approximately 100 students, the chance of misadventure increased.
5. **Absenteeism**: In addition to the above, high levels of absenteeism occurred as most visits had to be scheduled at times outside of timetabled teaching blocks. Students were often unwilling to attend visits outside of scheduled teaching times.
6. **Travel Distances**: For those students who did attend the scheduled site visits, travel distances and locations were also an issue. Many of the appropriate sites were not located at sufficiently central locations making attendance by public transport unfeasible, or involved excessive commuting times for students.

A compromise arrangement was in place immediately prior to the implementation of *OnSite*, in which students individually took on the responsibility of sourcing, contacting and liaising with a contractor and visiting a site in order to compose the visual diary. As a consequence of this amendment, the relative importance of the diary overall was diminished as part of the student assessment to reflect the release of governance of the course academic. While this model solved some of the issues surrounding the management and coordination of the site visit schedule, it inevitably brought about a diminishment of effort on the part of students to complete the task.

Situational eLearning

It is within this context that the authors participated in a research project, *Situational eLearning: A Crowdsourcing Approach to the Definition and Assessment of Key Practice-Ready Academic Outcomes* funded by the Australian Office for Learning and Teaching for the development of a number of eLearning resources to address the challenges numbered above (Newton, Hardie, Lowe, Pickersgill, Rameezdeen, Srivastava, Zillante, 2015). The principal investigator within that project, Professor Sid Newton from the Faculty of the Built Environment at UNSW, developed a research proposal that reflected the challenges above. In particular the research proposal addressed the need for blended learning strategies in the development of practice-ready professionals in the architecture, engineering and construction industries. The authors, as part of a team of academics from University of New South Wales, Adelaide University, Western Sydney University and the University of South Australia, contributed knowledge of VLEs and the development of appropriate Learning Platform strategies as a part of the project's overall aim.

The project investigated the possibilities for multiple platforms for the teaching and monitoring of learning patterns within Architecture and Construction course at a tertiary level. A core premise of the larger project asked how to address the educational requirement for AEC students that they should have some first-hand knowledge of the 'reality' of how a building site functions, site visits being considered a significant component of the 'practice-ready' graduate (Fletcher, Meyer, Anderson, Johnston, Rees, 2012). The reality that contemporary class sizes within introductory classes generally exceeded 100 students, particularly in Sydney and Penrith, meant that many of the core presumptions of a successful educational module requires consistency of access and delivery to the course content. Logistically, the task of ensuring that students are all able to access the same site, within a reasonable time frame of its development, was proving impossible. Further, the logistics of students actually taking the responsibility of attending external classes often meant that there was considerable absenteeism for this component, as detailed above.

OnSite Applied

The response of the UniSA team to this challenge was to develop a VLE that included a GE-based set of site visit environments that captured five different stages of the construction process: Site setout; slab; single-storey wall framing; double-storey wall framing; roof framing. The name chosen for the project overall was ***OnSite*** to reflect the emphasis upon immersivity and allude to the idea that the participant was going 'on site' and was simultaneously 'situated' within the digital environment.

OnSite is an immersive learning environment (VLE) that employs a commercial game engine (Unreal Engine 4 or UE4) to create a realistic digital environment that allows students to move around a domestic construction site using the same type of interactive commands consistent with a commercial video game (Dicheva, Dicheva, Agre, Angelova, 2015). In contemporary game environments, technical developments in the reproduction of 'realistic' formal, material and lighting aspects of environments has been an acknowledged aim of commercial developers and publishers of games. 'Sand box' environments, a subset of game environments generally, are a form of immersive environment in which the player is situated in realistically-scaled locations, and in which they are free to roam over an area defined only by the boundary of the map. The term 'sand box' comes from the allusion to the child's play enclosure

Figure 2. OnSite slab environment (UniSA, 2017)
Source: UniSA 2017

in which children are encouraged to create structures and games within a defined area, but with an open series of outcomes based on the nature of spontaneous play. While *OnSite* is not strictly a full sand box environment as it was limited to a single participant, it was still premised on the opportunity for the student to roam freely around the site, inspecting the progress of construction.

For *OnSite* the authors designed a limited sand-box environment at the scale of a domestic housing lot, in which the student could observe a stage of the construction process. Where a commercial game encourages the player to interact with the environment to achieve tasks relevant to the gameplay, *OnSite* created a sequence of site conditions that were typical of site-inspection stages as listed above (Figure 2, Figure 3).

Figure 3. OnSite slab environment (UniSA, 2017)
Source: UniSA 2017

It is important to distinguish, at this point, that while the *OnSite* project was undertaken using the game-engine software native to the UE4 platform, and it was not designed to be a 'game' in the sense of being game-based learning. While there is certainly scope for the development of these types of learning environments within the *OnSite* asset, at the outset it functioned as a didactic environment rather than one that required specific types of interactivity. Students were encouraged to explore a static environment for the purposes of checking its compliance with the other media of communication they had access to. In addition, it required some observational skill on the part of the student to recall anomalies in the built work in response to the on-line quizzes developed to test their experience. A comparison with a contemporary form of game-based learning is instructive to distinguish the difference between contemporary emphases on game-based learning that incidentally employ 'simulated' environments such as *OnSite*, versus the deep investment in visual fidelity the *OnSite* employs.

For example, Carina González and Francisco Blanco (González & Blanco, 2008) employ the game environment *Neverwinter Nights* as a platform for integrating secondary school mathematics with a Moodle site that tracks progress in completing tasks. The quest-based environment encourages students to progress through the game space by completing tasks associated with the mathematics curriculum, achieving academic progress while entertaining and motivating students to improve their skills. The inclusion of the Moodle platform ensures there are appropriate metrics to monitor student progress and integrate relevant educational goals. In terms of the goals of promoting collaborative learning and social interaction, González and Blanco's model works perfectly, but it is worth distinguishing between the level of fidelity and visual authenticity achieved in that game versus *OnSite*. For their purposes the 'fantasy' environment and low-poly characters, texturing and lighting were irrelevant for the principal aims. Arguably, an investment in a more 'realistic' modality would have diminished the entertainment value of the *Neverwinter Nights* environment. By contrast, for *OnSite*, the need for realism and immersivity directly influenced the approach to authenticity that framed the pedagogical approach.

Reviewing the considerable literature emerging from journals such as *Simulation and Gaming*, it is clear that there are a number of lexical distinctions in the field that help to clarify key research questions and to compare cognate explorations of the viability for game-based learning. Broadly speaking, 'simulation' carries two homonymic meanings, that a game is *inter alia* a simulation of a set of affairs that can be reiteratively played out to test multiple outcomes, while it can also be a simulation in which its fidelity to a set of standards is the purpose of the test – how 'real and accurate' a simulation is the experience. In the first instance, the issue of fidelity to the real is trivial in comparison with the pattern of play and outcomes explored; in the second fidelity is the goal of the process. Arguably, the issue of fidelity to

Figure 4. OnSite two story framing environment, (UniSA, 2017)
Source: UniSA, 2017

the real is also a distracting factor in the first iteration of simulation. A chess game, for instance, in which the pieces are entirely life-like arguably has little influence of the strategy and goals of the game of chess itself. Indeed, and acknowledging Lewis Carroll, the more protean and elaborate the narrative that may be constructed around the pieces, the less stable (and perhaps less whimsically interesting) is the quality of the game-play.

And yet it was clear, anecdotally, that students considered the synthetic nature of the *OnSite* environment to be one of simulation, free from the constraints of real-world interactions. However, it was also the case that the authenticity of the

Figure 5. OnSite two story framing environment, (UniSA, 2017)
Source: UniSA, 2017

environment was a factor in determining its validity as a teaching and learning experience (Niculae & Duda, 2015). So while *OnSite* was not a game, in ludological terms, nor did it require social constructivist behavior within the environment, it still referenced the greater category of game-based learning because of the clear use of simulation and reiterative experience.

OnSite was premised on a number of technical and psychological assumptions regarding the capabilities of a GE environment to deliver enhanced levels of perceived fidelity (Figure 4, Figure 5), experiences aimed at consistency with reality but with also the tacit understanding that this was also a focused and artificial experience that required the students to examine or be conscious of it as a learning environment. For the reason that it is possible to undertake a site visit and the information exchange to be merely part of the continuum of experience, *OnSite* attempted to make its 'realism' a confirmation of the legitimacy of the information it contained.

In addition, the negotiable environments were embedded within a Moodle learning plan that encouraged the exploration of the material though a variety of learning processes. Both the immersive environment and the Moodle site worked together to deliver a focussed learning experience.

Context and Methods

The project originally was agreed upon on the following conditions, specifically ease of use, staff specialization, technical support and learning curve for student expertise.

1. Students successfully downloaded and ran the VLE software for *OnSite* across a number of personal learning devices, as well as devices available in UniSA learning centres.
2. Minimal staff intervention was necessary to solve technical issues regarding the access and use of the *OnSite* VLEs. Forum questions were minimal over the three-year period.

Teaching staff were not originally involved in the technical development of the *OnSite* material in 2015. Subsequent familiarity with the process in 2016 and 2017 assisted teaching staff in responding to student queries on an ad hoc basis, though again interaction was generally not necessary as students cooperatively assisted each other. This fact reinforced the social constructivist aim of the project overall.

Pedagogical Methods

Chris Dede summarized the potential for future research into immersive teaching environments by making four points that reflect a number of the outcomes of work completed by himself and associated researchers, and which are relevant to the *OnSite* project (Dede, 2009). These are:

1. To what extent does good instructional design for immersive environments vary depending on the subject matter taught or on the characteristics of the learner? For what types of curricular material is full sensory immersion important?
2. To what extent can the successes of one's virtual identity in immersive environments induce greater self-efficacy and educational progress in the real world?
3. To attain transfer, what is the optimal blend of situated learning in real, augmented, and virtual settings?
4. What insights about bi-centric frames of reference can generalize from immersive environments to pedagogical strategies in face-to-face settings?

Whilst Dede's questions refer generally to the total enterprise of immersive educational environments, it is clear that the structure and outcome of the *OnSite* project addresses and matches the core ambition of the questions: to enhance learning outcomes through immersivity in synthetic and augmented environments. The *OnSite* model, as a component of the *Situational eLearning* research project, demonstrates that the investment in simulated environments for introductory material has a great deal of potential in terms of pedagogical development in the architecture and construction industry. By paying attention to the needs for authentic learning experiences that emulate the complex conditions of construction, and the development of a shared repository that is relevant and accessible to construction educators, there is a clear proof of concept that radical innovation in this domain of learning is possible and achievable.

In addition, when employing digital environments such as *OnSite*, there is considerable literature on the effectiveness of immersive software and its relationship to an experience of authentic learning. In these conditions Jan Herrington and Jenni Parker (Herrington and Parker, 2013) identify that authentic learning using emergent technologies requires a number of conditions to be met:

1. Authentic context
2. Authentic tasks and activities
3. Access to Expert performances
4. Multiple perspectives
5. Collaboration
6. Reflection
7. Articulation
8. Scaffolding
9. Authentic assessment

Figure 6. OnSite Moodle access incorporated as part of the course home page, (UniSA, 2017)
Source: UniSA, 2017

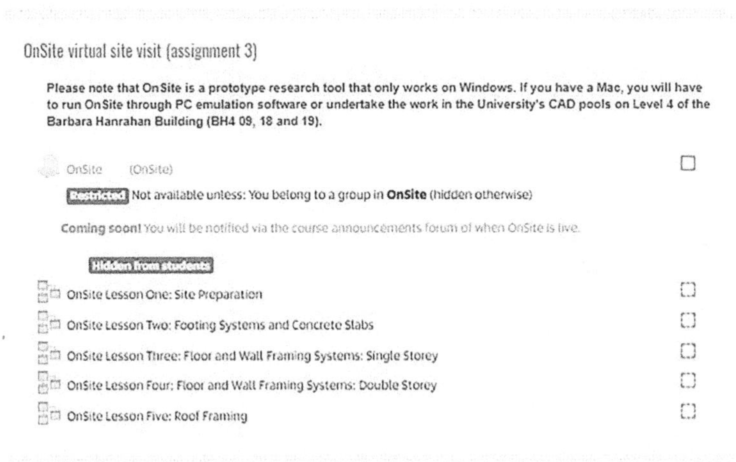

OnSite addressed this issue by creating a site that was, by definition, consistent for all students, accessible, appropriate in terms of learning objectives for construction content and sequencing. In addition, a dedicated Moodle Learning Plan was created that required students to view the material in a particular sequence consistent with a self-directed learning programme (Figure 6).

Within the Moodle plan, students were initially directed to an external website repository, SeLAR (Situational eLearning Repository) that contained the necessary files for the *OnSite* environment as well as a variety of relevant documents (plans, sections, etc) and images that described the project context. Once these files were downloaded, the students were able to run the game file on their PC, or via Apple Boot Camp, to enable them to be within the environment.

After this material was accessed, students were required to view a 6-10 minute YouTube video in which the site in question was navigated by an instructor, pointing out relevant details consistent with a site inspection that would be conducted within a real-world scenario (Figure 7). Following the *OnSite* experience and the YouTube video, students were asked 10 multiple choice questions on material relevant to the site visit and the knowledge base that it demonstrated. At the end of each question, students were given an indication whether their answer was correct and prompted to proceed to the next question. At the end of the 10-question quiz, students were offered the opportunity to repeat the quiz in order to improve their grade. If they chose to do so, the questions remained the same, but the location of the correct multiple-choice answer changed requiring students to concentrate on and reinforce

Figure 7. OnSite short instructional video for the Site Establishment VLE (UniSA, 2017)
Source: UniSA, 2017

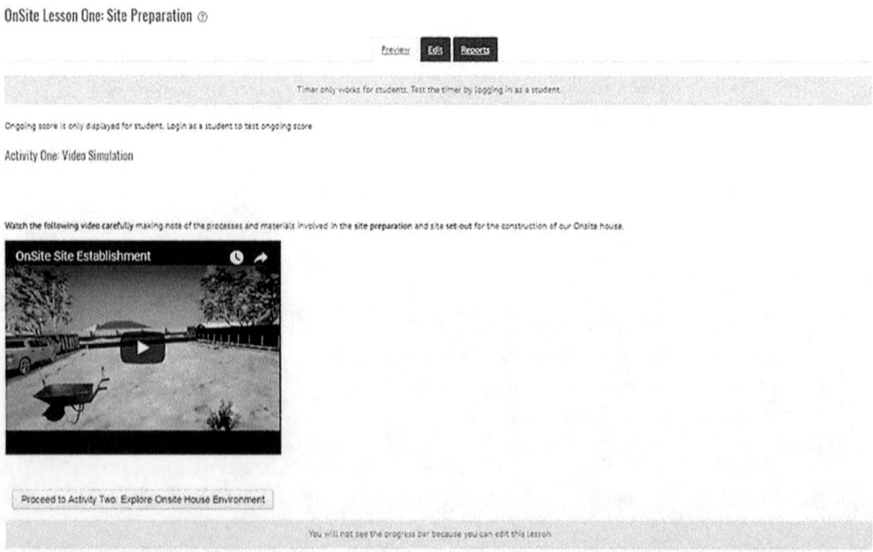

the correct response. Students were given a window of 30 minutes to complete the quiz, assuming they would switch between the *OnSite* environment, the supporting documentation and the relevant YouTube video to find the correct answer. This pedagogical framework was consistent for the 5 separate *OnSite* environments (Figure 8).

Following Herrington and Parker's principles, it is possible to analyse the effectiveness of the *OnSite* Project and to reflect on the alignment between the project's initial intentions and its evolution over three years of use.

Figure 8. Design Construction Principles: OnSite Student workflow (UniSA, 2017)

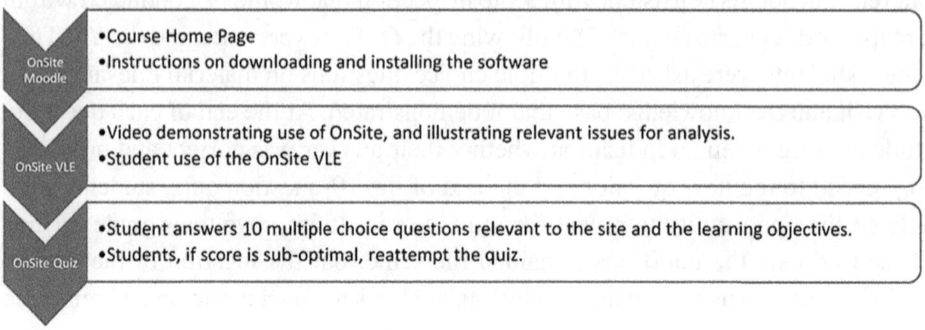

1. **Authentic Context:** Herrington and Parker identify 'authenticity of context' as a key pre-condition for a successful pedagogical framework in which self-directed and e-learning tools are considered to be appropriate for the learning outcomes of the course and, importantly, are considered relevant to the course material as well. In the case of *OnSite* there was an implicit endorsement of the authenticity of the project because of the fidelity to a 'real world' example (Figure 8), and its incorporation into the summative assessment attached to the learning process. In addition, the course itself focusses upon acquiring familiarity with a specific body of knowledge, with little scope for a discretionary opinion on its value by the student.
2. **Authentic Tasks and Activities**: As described above, the relevance for the course to the preparation of a 'practice ready' student made this condition implicit. While the final outcomes of how different student types might ultimately employ the knowledge of domestic construction may have differed between architecture and building students, the confirmation that this material represented core knowledge was recognized in the course descriptors and reinforced verbally by teaching staff.
3. **Access To Expert Performances:** As part of the suite of information created to support the independent employment of the *OnSite* material, students were provided with a step-by-step guide to downloading, installing and running the software (itself a positive outcome), which was developed through access to on-line videos that demonstrated how the 5 different levels could be experienced. In addition, these 'expert' videos critically scrutinized the environments and drew attention to features that would later form part of the assessment rubric.
4. **Multiple Perspectives:** The context for undertaking the *OnSite* project allowed students to complete the task either at home on their personal devices, or in University Computer Pools in which co-learning had been established as an appropriate form of task completion. In addition, official forums and

Figure 9. OnSite single story VLE (UniSA, 2017)
Source: UniSA, 2017

unofficial messaging platforms were employed to allow students to discuss aspect of the *OnSite* project. In this way, access to multiple perspectives was expected though assessment was wholly individual. An initial concern from teaching staff regarding student 'collusion' was addressed by discussing the relationship between the acquiring an understanding of course content, student co-participation and the assessment method. Importantly for the *OnSite* project, it was agreed that the encouragement of students to discuss the material and to make multiple attempts to attain a high grade actually served the purpose of reinforcing the material and ensuring persistent knowledge through repetition and reinforcement.

5. **Collaboration:** As above, the capacity for mutual discussion and collaborative effort was recognized as an important part of the learning process and the assessment devices. Requiring students to complete 5 quizzes, each with 10 multiple choice options, meant that there were a number of discussion points that could be discussed and decided upon as questions of fact.
6. **Reflection**: Course evaluation Instruments were included for each of the iterations of the Construction course, covering the entirety of the course content as well as *OnSite*. Reported feedback included positive support for the *OnSite* project, however it was also noted that ultimately a 'real life' visit to a site would have been valuable.
7. **Articulation:** As demonstrated earlier, the course materials supported the student exploration of the VLE. Reviewing the lecture material, exemplar material and differing course documentation media allowed students to recognize the core purpose of the VLE itself.
8. **Scaffolding:** The *OnSite* project constituted a component of a multi-modal form of construction education, designed to introduce students to an authentic site that simulated authentic construction conditions (Figure 9). One of the avowed intentions of the project was to produce a digital model that was accurate in terms of scale, material and constructional method. Ensuring this level of authenticity prepared the students for the scrutiny of more complex construction environments, digital and real, and introduced them to a method for assessing the relationship between the media of communication used in contract documentation and real-life scenarios.
9. **Authentic Assessment:** Assessment was based on a number of aspects of the VLE. Principally students were asked to review certain aspects of the site and recollect whether the VLE, as modelled, reflected standard, correct industry practice (Were the footings correctly set out? Was the timber studwork correctly located? Etc.). Recognizing successful and unsuccessful compliance with standards is a core responsibility of a future professional tasks in the construction environment, hence the questions and the assessment reflected

that requirement (Amory, 2012). Overall, the better the student could recognize what satisfied a correct standard for a safe, well-organized and appropriately sequenced construction site, the better they were able to perform within the assessment. For this reason the assessment itself was recognized to be appropriately authentic.

Results from the employment of the five *OnSite* VLEs over three 13-week semesters, 2015-2017, indicate a number of outcomes that support the initial pedagogical aim to ensure that material covered in the course description was recognized and engaged with in the learning process. The *OnSite* material was intended to give students access to a select range of site conditions relevant to the overall course aims. The *OnSite* quiz results constituted 20% of the overall semester grade. Students completed 5 quizzes with a maximum of 10/10 points per quiz, therefore each individual quiz constituted 4% of the overall semester grade.

Summarizing the student experience, the following can be noted:

1. Students successfully engaged with the SeLAR repository website to review relevant materials;
2. Students had access to UniSA Computer Pools in which they could download and run the VLE software for *OnSite*;
3. Many students successfully downloaded and ran *OnSite* across a number of personal learning devices (PC only, though Mac use was possible if they were running a Windows OS).
4. Minimal staff intervention was necessary to solve technical issues regarding the access and use of the *OnSite* VLEs. Forum questions were minimal over the three-year period.

In addition, while course teaching staff were not originally involved in the technical development of the *OnSite* material in 2015, subsequent familiarity with the process in 2016 and 2017 assisted teaching staff in responding to student queries on an ad hoc basis, though again interaction was generally not necessary as students cooperatively assisted each other. This fact reinforced the social constructivist aim of the project overall.

Analysis of Outcomes

Amongst other outcomes, we note the following:

1. Students successfully completed the quizzes employing the blended elements of reference materials, video tutorials and personal observation.

Figure 10. OnSite roof framing VLE (UniSA, 2017)
Source: UniSA, 2017

2. Many students undertook multiple attempts at the quiz element to improve their overall grade. (See Table 1)
3. From Outcome 2 above, it can be construed that despite the fact that a student might achieve a score of 8/10 versus 10/10 (say) which would have minimal impact on their overall grade, students nevertheless returned to the material at a significant rate if their grade was less than optimal. This was considered a successful outcome of the project.
4. Reviewing the distribution of *OnSite* grades and overall performance generally indicates that student success overall was consistent with effort within the *OnSite* component (Figures 12-13). The x axis number indicates an individual student identifier, while the y axis charts the percentage grade outcome for the *OnSite* component and the course grade overall.
5. Area charts for 2015-2017 (Figures 14-16 below) indicate significant success in students applying the correct answer to questions. Responses in the lowest percentile were from students who did not engage with the *OnSite* at all or minimally.
6. Students initially undertook significantly more attempts at achieving a suitable grade for the initial VLE: Site Setout. Having acquired a familiarity with the process, the average attempt reduced significantly.
    a. Micro-analysis of the individual question responses indicated in the final VLE: Roof Framing, infer that incrementally subtler and more difficult choices were being asked for, hence the increase in average quiz attempts generally.
    b. Students that achieved 100% tended not to repeat attempts to improve their grade or to review their responses. For this reason, the motivation to improve was even stronger for those not achieving the maximum mark.

Figure 11. User results, overall course grade and OnSite, 2015 (UniSA, 2015)

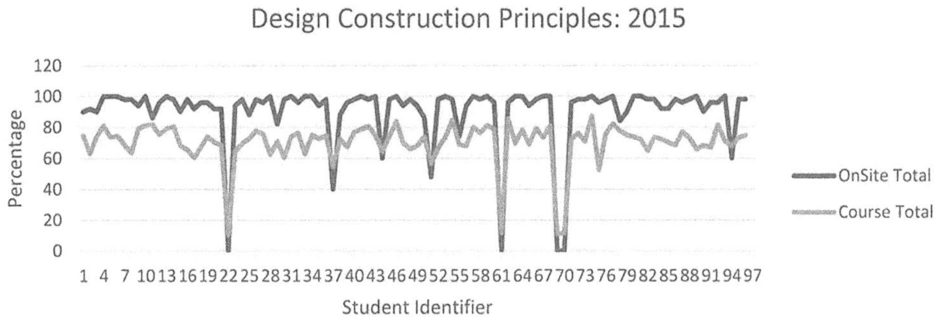

Student Experience and Behaviour Data

The participation data below tracks the relationship between an individual student's final grade for the *OnSite* component of the course, and their overall grade. As can be seen, there is a strong indication that *OnSite* grades assisted students in achieving an improved overall grade. This was a recognized and predicted outcome of the exercise as the general pedagogical aim was to encourage repetitive engagement in an effort to earn the highest grade possible. Other assessments were designed, and assessed, on the principal of a bell-curve grading system.

Student Identifier Numbers on the x-axis refer to individual students within the *OnSite* course component and the course overall; y-axis numbers refer to the *OnSite* component and overall course grade for students, expressed as a percentage.

Table 1 analyses data regarding multiple attempts at the *OnSite* quizzes. It tabulates the average number of multiple choice quiz attempts per VLE. Columns detail the difference between the raw average of the number of attempts at the multiple-choice quiz (A columns) and the attempts by those students who made multiple attempts

Figure 12. User results, overall course grade and OnSite, 2016 (UniSA, 2016)

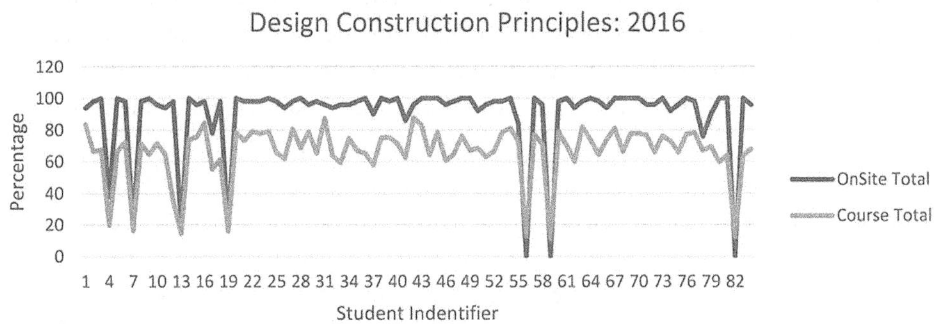

Figure 13. User results, overall course grade and OnSite, 2017 (UniSA, 2017)

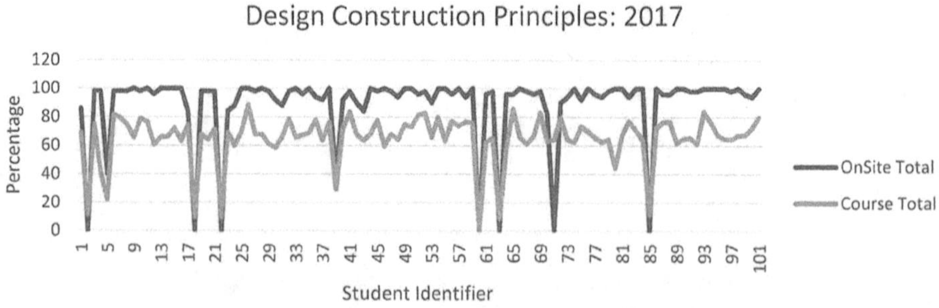

Table 1. Graph showing all user attempts and multi-user attempts, (UniSA, 2017)

Year	2015		2016		2017	
VLE Identifier	**All Attempts User Avge.**	**Multi-Attempt User Avge.**	**All Attempts User Avge.**	**Multi-Attempt User Avge.**	**All Attempts User Avge.**	**Multi-Attempt User Avge.**
	A	B	A	B	A	B
Site Setout	4.772	5.232	4.247	4.528	3.494	3.689
Slab	2.424	3.079	2.456	2.966	2.400	3.030
Single Storey Framing	2.143	2.733	1.846	2.535	1.851	2.558
Double Storey Framing	2.022	2.765	1.870	2.595	1.815	2.471
Roofing	1.854	2.490	2.156	2.816	2.547	3.100

(B columns) to improve their grade. B column students had multiple exposure to course material. Roughly speaking, most multi-attempt students made 2-3 more attempts to improve their grade.

Area Charts for *OnSite*

In addition to the Overall Grade and Multiple Attempts data, we also calculated and tabulated the Area Charts for all students for the *OnSite* component. As the project was designed to, ideally, encourage all students to achieve a perfect score, the larger the shaded area the higher the overall grade result for the student cohort.

REFLECTION

Reviewing the statistics above, we can make a number of observations relevant to methodology, content, student behavior and future directions (Lu & Law, 2012).

Figure 14. OnSite area chart, 2015 (UniSA, 2015)

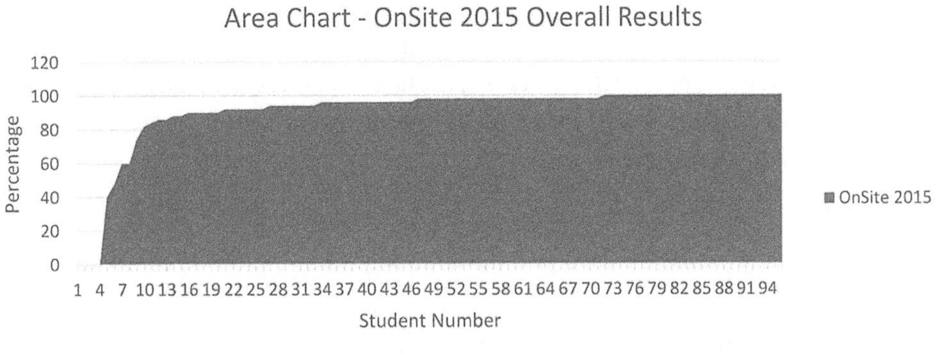

Figure 15. OnSite area chart, 2016 (UniSA, 2016)

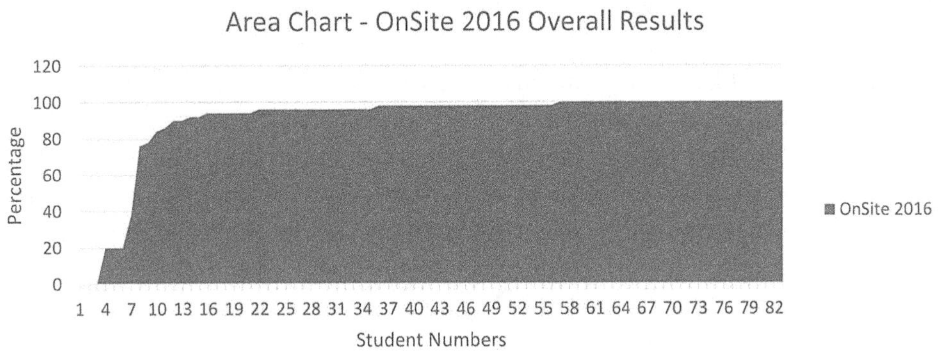

Figure 16. OnSite area chart, 2017 (UniSA, 2017)

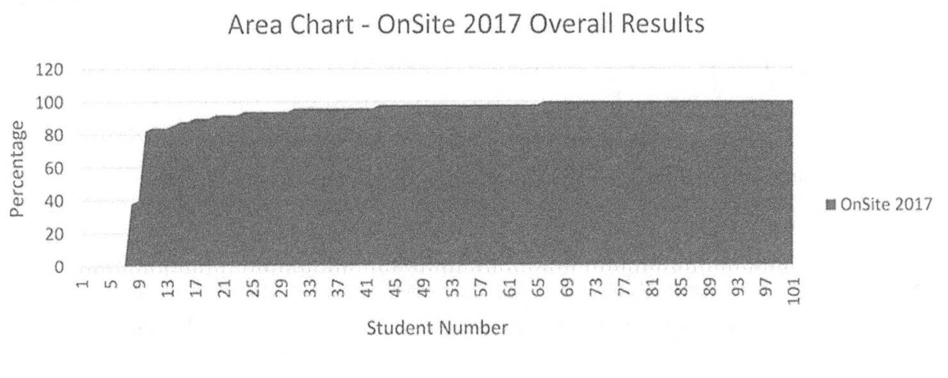

When creating the original *OnSite* the authors, and our research colleagues, created environments and designed the types of site visit 'experiences' we felt were relevant to an introductory construction course in architecture. While there was considerable confidence that there would be engagement with the material, and that technical issues of compatibility with personal devices would be minimal, we were pleasantly surprised that these issues were only minor in number and complexity. Having employed the *OnSite* project for three successive years, the data now shows that there are consistencies of approach and engagement that demonstrate a positive effect on the student experience.

SUMMARY

In summary, we are of the opinion that the *OnSite* project has successfully developed and incorporated the educational content relevant to the course Design Construction: Principles. As a test case for introductory constructional knowledge across the Australian design and construction tertiary environment, the tool has considerable applicability provided it is adequately implemented into course assessment (Albeanu, 2015). However, like most game-based learning strategies, it requires ongoing management of the resource to ensure that there is a seamless encounter by the student in the context of their learning platform. Trials at Western Sydney University and other courses within UniSA had more limited success because of the tool's incidental incorporation into course management. We are mindful of the need to provide expertise and support to other trials of the *OnSite* material.

From the project itself a number of key questions remain, that can successfully evolve from the *OnSite* project. These are:

1. Does an increased level of immersion in a virtual learning environment imply (or require) a greater sense of the 'reality' of the location and the authenticity of the simulation?;
2. Does an increased level of immersion effectively mean a greater degree of knowledge acquisition?, and;
3. Does an increased level of immersion definitively lead to improved educational outcomes?

Future research directions for the *OnSite* project will involve the development of a multi-user environment that will allow for more specifically social-constructivist behavior to take place as part of the learning and assessment rubric. Allowing multiple users to cohabit a shared environment in real time, including tutors and assessors, would fundamentally change the nature of the educational experience. Importantly, the shared space would allow the *OnSite* environment to evolve from one directed at best-practice and defect identification, to an environment that more realistically depicted a site meeting between various stakeholders in a construction project. The ability of students to role-play the responsibilities of contractor, sub-contractor, client, architect, planning officer, etc would encourage significantly more subtlety in the learning experience and permit greater granularity in discussing issues associated with the construction process.

REFERENCES

Albeanu, G. (2015). Using Moodle for Teaching Science and Information Technology, Towards Higher Education Reengineering. *11th International Scientific Conference, eLearning and Software for Education*, 606-611.

Amory, A. (2012). Tool-Mediated Authentic Learning in an Educational Technology Course: A Design-Based Innovation. *Interactive Learning Environments*, *22*(4), 497–513. doi:10.1080/10494820.2012.682584

Dicheva, D., Dicheva, C., Agre, G., & Angelova, G. (2015). Gamification in Education: A Systematic Mapping Study. *Journal of Educational Technology & Society*, *18*(3), 75–88.

González, C., & Blanco, F. (2008). Integrating an Educational 3D Game in Moodle. *Simulation & Gaming*, *39*(3), 399–413. doi:10.1177/1046878108319585

Herrington, J., & Parker, J. (2013). Emerging Technologies as Cognitive Tools for Authentic Learning. *British Journal of Educational Technology*, *44*(4), 607–615. doi:10.1111/bjet.12048

Lu, J., & Law, N. (2012). Understanding Collaborative Learning Behaviour from Moodle Log Data. *Interactive Learning Environments*, *20*(5), 451–466. doi:10.1080/10494820.2010.529817

Maghool, Hossein (Iradj) Moeini, & Arefazar. (2018). An educational application based on virtual reality technology for learning architectural details: challenges and benefits. *International Journal of Architectural Research, 12*(3), 246-272.

Newton, S., Hardie, M., Lowe, R., Pickersgill, S., Rameezdeen, R., Srivastava, A., & Zillante, G. (2015). Situational eLearning: the new potential for digital technology. In RICS COBRA AUBEA 2015: The Construction, Building and Real Estate Research Conference of the Royal Institution of Chartered Surveyors. The Australasian Universities' Building Educators Association Conference.

Niculae, R., & Duda, M. (2015). Gamification as a Means of Getting from e-Learning to Experience Learning in Architecture. Carol I' National Defence University.

Savage, S., Davis, R., & Miller, E. (2010). *Professional Education in Built Environment and Design*. Australian Learning and Teaching Council.

Tucker, R., Abbasi, N., Thorpe, G., Ostwald, M., Williams, A., Wallis, L., & Kashuk, S. (2014). *Enhancing and assessing group and team learning in architecture and related design contexts*. Office for Learning and Teaching.

Zheng, L. (2015). A systematic literature review of design-based research from 2004 to 2013. *Journal of Computer Education*, *2*, 399. doi:10.100740692-015-0036-z

Chapter 10
Visual Tools for Analyzing and Organizing Engineering Data and Information

Felix Kin Peng Hui
The University of Melbourne, Australia

Hamzeh Zarei
The University of Melbourne, Australia

Colin Duffield
The University of Melbourne, Australia

ABSTRACT

The chapter examines the use of visual tools in inquiry-based learning in engineering education. Engineering information has traditionally been communicated using engineering drawings. The advancement of technology has empowered engineers to work with data to generate graphics where visual tools have been successfully employed in the industry. However, many educational institutions still rely on text-based learning, which rely on linguistic abilities to convey complex concepts. Using program logic theory, the authors evaluate how a poster can help learners acquire knowledge and know-how to represent complex ideas in a logical and consistent manner. Students reported benefits in poster exercise compared with a written assignment. This chapter also discusses the educational and managerial significance of using visual tools to greater effectiveness and the importance of communicating technical information.

DOI: 10.4018/978-1-5225-8452-0.ch010

INTRODUCTION

Engineering information is conveyed in many ways. Traditionally, technical drawings or engineering graphics are used to convey information, ideas and facilitate visualization of the final product. For years, the teaching engineering graphics has been progressed from hand-drawn sketches and drawings to computer-aided design using software packages capable of solid modelling. Software packages are now capable of producing 3-D models with animation. Along with advancement in computing technology, the data used in the engineering drawings are now reliably linked with to management functions in construction and fabrication such as like scheduling and risk analysis. Apart from conveying information in drawings, engineers also required to present complex information in many other ways including the use of graphs, charts or process diagrams.

Problem solving and providing solutions are a big part of an engineer's skillset. However, in providing solutions, it should be noted that real-life engineering problems are part of bigger problems, some of which may not just technical. Ackoff (1974) noted no problem exists in isolation and every problem is a part of a system of problems. Providing solutions to engineering problems require interaction with and within social systems that encompass internal and external stakeholders. Problems encountered in these complex systems can be "ill-structured", "messy" or involving "social messes". The term "wicked problem" (Ackoff, 1974; Churchman, 1967; Horn, 2001) is used to describe an issue in systems that are complex, ambiguous, severely constrained, interconnected, are seen differently and even contradictory from different points of views (Rittel & Webber 1972; Simon 1984). Wicked problems are multi-facets and often change over time (Churchman, 1967).

Due to methodological limitations or data impediments, engineers and managers face challenges in incorporating their observations of all phenomena. Challenges include gathering data from stakeholders, consolidating and integrating data and making meaningful representations of these data. Therefore, finding a better way to consolidate data from observations, finding better ways of identifying any underlying patterns and synthesizing related data using analytical tools to identify root causes will lead to improvements and solutions.

Challenges in Communicating Technical Information and Opportunities

As technology advances, new approaches towards the visualization of data should be included in engineering graphics curricula. Although graphs, charts and diagrams are taught and used in most engineering subjects, there is really a need to revisit these visual methods of presenting engineering data and incorporate these as standard engineering graphics training that can help engineers understand the importance of engineering data and visualization process. This book chapter revisits some of the ways of presenting engineering data that most educators take for granted and present them in a concise manner that can be used to enhance the importance of teaching visualization. In particular, the poster is a useful tool that can be used to combine all these charts and data and present them in a concise manner to the audience. The poster as a tool integrates other visual tools and synthesizes related and concepts. The poster can also be developed quite quickly with personal-computer based tools such as Microsoft Powerpoint or Microsoft Word. It is the aim of this chapter to illustrate the use of poster as a tool to train engineers to overcome these challenges associated with complex idas and to present complex data in a meaningful way.

Methodology

In this chapter, the methodology used to showcase the use of poster in training engineers to handle complex data will the program logic theory (Funnel and Rogers 2011). Program logic theory is used extensively in government program and education research evaluation (NSW Government 2016 and CESE n.d.). A literature review on visual tools used in engineering is first carried out. The review looks at nature of complex information and how it can be organized in a visual manner. It will also look at some common visual tools used by businesses to convey technical and engineering information. The review will also cover some common visual tools that are used to integrate and synthesis information such as the poster or a one-pager. The chapter then showcases the use of a poster exercise in project management class and examines some of the education research that has been done on the use of the poster to present engineering information. The chapter describes how students are then asked to compare the poster exercise to a written assignment, a more conventional assessment method. The findings are presented in a detailed discussion of the how the poster exercise has aided learning. The educational and managerial significance are also discussed, this chapter aims to show that apart from an engineer's skillsets in computing, programming, cognitive skills in mathematics and problem-solving, it is also important to develop skills to produce and present this information in a convincing manner.

A REVIEW OF VISUAL TOOLS

In this section, we review some of the visual tools that have been effectively used in industry to encourage corporate learning and information dissemination. These are fast becoming important as the data can be drawn from the same dataset used for producing engineering drawings e.g. parameters used in drawings are also be used Computer aided design/Computer-aided manufacturing (CAD/CAM) applications or scheduling/planning applications in Building Information Modelling (BIM) systems. Visual tools also play an important role in data presentation required for managing engineering systems and decision-making. Most of these can be adapted as learning tools in an educational setting. In essence, most of these tools require the learner to put together ideas, construct solutions and present them as they advance. Visual tools have three main characteristics:

1. Level of abstraction of a visual artefact may determine the quantity and quality of data, information or knowledge it contains. Subjective tools may accommodate abstract concepts or illustrate a concept model through representation of visual elements. Objective tools on the other hand build a presentation in an efficient and easy to understand platform. Factual tools may suit large or assorted data sets that requires presentation in a holistic and concise format.
2. Assembly of a visual tool might be structured, semi-structured or non-structured. It defines the flexibility and traceability of visual artefacts in accommodating assorted information. In a structured tool, information will always generate same visual presentation. Little or no flexibility exists for the author to influence the final representation of data. On the other hand, non-structured visual tool means that a same information might be visualised differently. Structured tools on the hand provide a level of traceability from the visual illustration to the original data. Semi-structured tools offer some extent of flexibility and traceability at the same time.
3. Inter-connection is another trait of a visual representation that concerns the inter-connection of data within a visual illustration to be relational or dimensional. While relational tools imply an association among visual data elements, the dimensional tools stretch a continuum for comparison among data.
   The proper application of visual tools of in engineering depends on the understanding of the capacity of the visual tool to accommodate and adopt the data it contains. This relationship is shown in Figure 1. The complexity of the data is a principal criterion for the selection of the proper medium to present data.

Figure 2 shows some of the common visual tools that are commonly used in engineering. These tools will be described in greater detail in the following section.

Table 1 shows the level of abstractness, assembly and inter-connection for some of the common visual tools.

Figure 1.

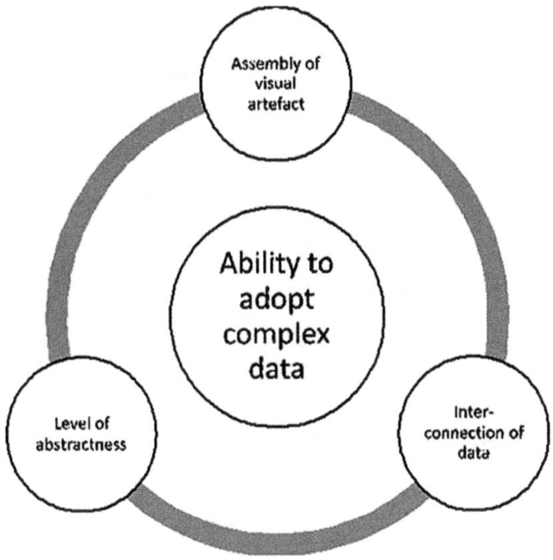

Figure 2.

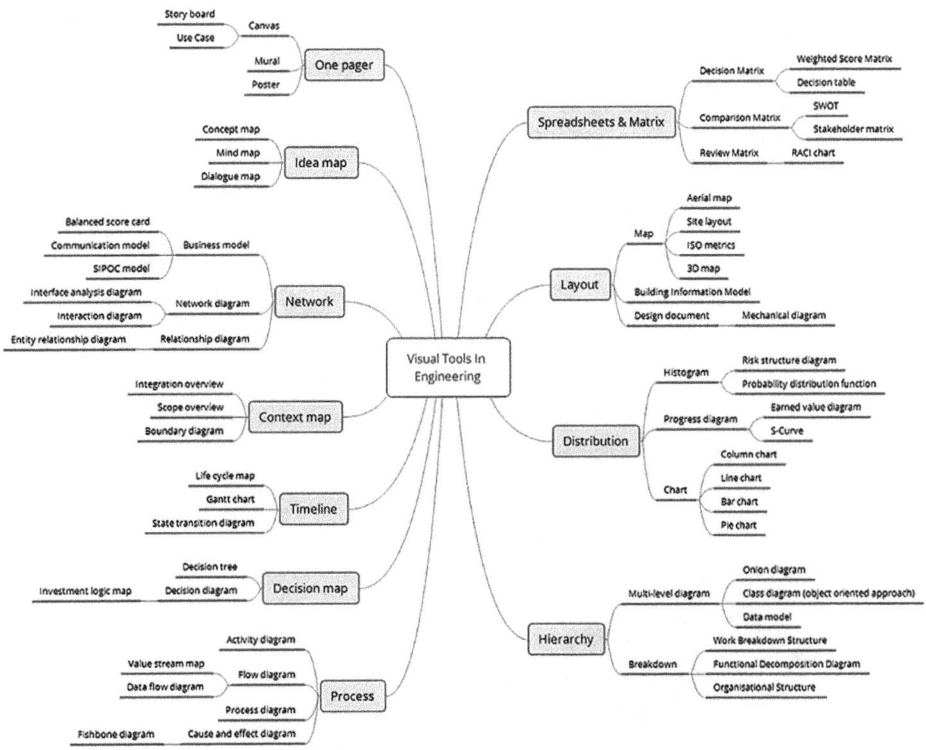

Table 1.

Visual tools	Abstractness	Assembly	Inter-connection
Spreadsheets	Various	Structured	Often dimensional
Layouts	Often objective	Structured	Often relational
Distribution tools	Often objective	Structured	Often dimensional
Charts	Often objective	Structured	Often dimensional
Hierarchy tools	Various	Semi-Structured	Often dimensional
Process tools	Various	Semi-Structured	Often dimensional
Decision maps	Various	Semi-Structured	Often relational
Timelines	Often objective	Semi-Structured	Often dimensional
Context maps	Often subjective	Non-structured	Often relational
Network tools	Often subjective	Non-structured	Often relational
Idea maps	Often subjective	Non-structured	Often relational
One pager	Various	Non-structured	Often relational

Graphs and Charts: Pareto Charts

Charts and graphs visually illustrate the relationships between numbers or numerical data (Lieu & Sorby, 2009). It is an intuitive way of showing relationships without the use of words or having the reader to analyse the raw data. In its simplest form, graphs translate numerical data into visual graphics that can show trends and patterns. Graphs represent data by using fixed locations on a flat plane with coordinate axes (Briscoe,1996). Figure 3 below shows some common graphs which includes pie graphs, bar graphs, histograms or frequency poly-gram, line graphs, scattergrams. Common commercially available software such as Microsoft Excel have features that can easily translated numerical data into graphs.

The bar chart is one of the most common forms of displaying numerical data (Lie & Sorby, 2009). One of the most commonly used forms of graphs in industry is the Pareto which is essentially a bar graph to show the frequency of incidences or occurrences. Pareto charts (shown in Figure 4) are commonly used as part of

Figure 3.

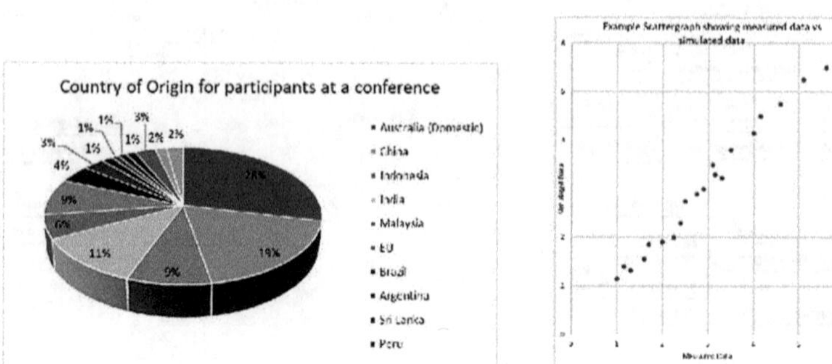

Figure 4.

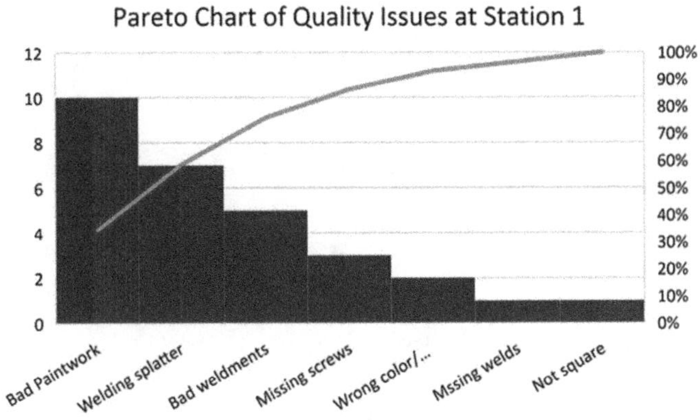

problem-solving techniques to force the thinking about 80/20 rule where 80% of the problems are caused by 20% of the problems (Juran 1974).

The line charts are also another common form of visual display for numerical data. The line chart is extremely useful for showing trends, or rate of change, or increases or decreases in numerical data.

The Pie chart is a useful way of showing how the proportions of parts that make up a whole are compared against each other. While scatter graphs or scatter plots shows the relationships between two variables plotted on the x and y axes and is best suited to illustrate data variability, data clustering and data trends.

In general, graphs need to be simple to achieve focus. Good use of colours, texture and graphical elements to emphasis the purpose of the graphs. However, Briscoe (1996) lists some of the common mistakes that prevent engineers from making clear and focused graphs. Figure 5 illustrates some of these mistakes that are common in poorly constructed graphs.

1. **Cluttering**: Too much information
2. **Misuse of Axis**: Axes that are too long (waste space), confusing increments (e.g. in increments of 3), correct use of offset axes (not starting at 0), excessive zero's making the graph hard to decipher
3. **Labels and Sizes**: Numbers being too large or too small
4. **Text**: Too small, or too big
5. **Legend**
6. **Symbols**
7. **Texture** and contrast for graphs e.g. inclusion of gridlines or not

Figure 5.

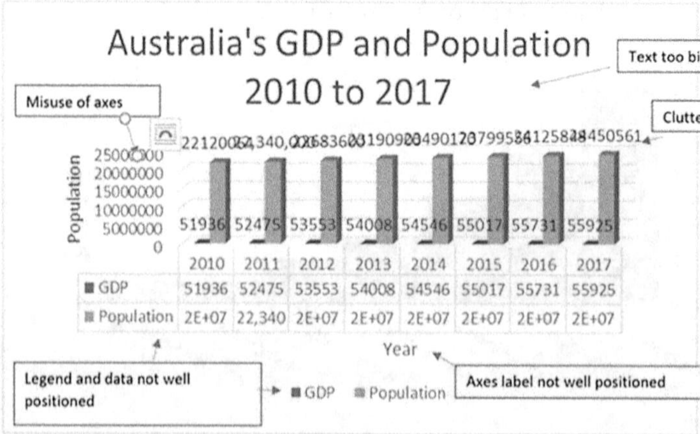

Diagrams

Process Maps and Layouts

Process flowcharts or process maps and Gantt charts are used in engineering to show how operational processes move in sequence. These provide a lot of information about how work is organised, how information flow, how it is communicated and how processes are controlled. Flowcharts are used extensively in engineering to show workflow and they display actions by people or machines. Flowcharts also known as flow maps show a precedence of time using block diagrams that present the flow of data or authority. Flow maps are useful in illustrating details of process and sequences of steps. Although flow maps can incorporate process details, the core structure is limited to contain station blocks and connecting arrows. Flow maps are easy to make and effective tools.

Gantt charts show the sequence of tasks on a timeline and is commonly used to describe schedules and sequence of events in projects. As it can be used to show relationships between activities, it is used a planning and control tool in engineering project management.

Business Diagrams

One important type of business diagram is an organisation chart. While organisation charts do not strictly contain engineering information, they add a lot of context to the situation surrounding the problem, more important it shows how human resources are organised in the context of the engineering issue or project.

Cause and Effect Diagrams or Fishbone Diagram

The fishbone diagram also known as the Ishikawa diagram has been widely used in problem solving situations. The basic idea to take a problem and analyse it in terms of groups of variables which can affect the process outcome. The strength lies in helping the user think about how the process can be affected by four or six different categories of variables such as Man, Machine, Materials, Measurement and Environment. Users or learners typically think through how these can various possible events or situation can cause a drift in the outcomes. Thinking visually in terms of variables can also simplify the problem. The sub-branches within each branch also help the user to visually move up the chain of effect to understand underlying root causes.

Mapping

Concept maps and mind maps are extremely useful tools for organising and presenting ideas. We will present some common ways of using these maps to present engineering-related applications.

A concept map is a presentation of data by identifying the concepts and the relationship between them. It is a semi-structured illustration, and an example of a concept map is shown in Figure 2. Concept maps establish a shared understanding between the author and audiences with a controlled emphasis on the link between author and audience. When applied to projects, they explore the conceptual framework of a problem and its boundaries. Concept maps are monologue unless they are prepared collectively. Concept maps have been extensively used in research, education, and learning processes (Jackson & Trochim, 2002). In projects, concept maps have been used to record a brainstorming session, display categories of various causes for an issue or self-reflect a personal understanding of an experience or a case. Nonetheless, the depth of information in a concept map is inadequate to illustrate a complex relation of different concepts or explain the root cause of a problem.

Mind maps are a representation of a thinking process in a visualised illustration; they are usually author focused. Mind maps are easy to make but not easy to read in a structured manner. They help record and review the flow of thinking and is highly dependent on the author's mindset. Mind maps might create bias since they are influenced by author's thinking process. The comparison between maps is hardly possible unless they are structured for comparison purposes. They are a monologue and usually only offer a one-way flow of information from author to audience. Mind maps are used to record individual reflections for saving time, connecting concepts, seeing the big picture, and helping memorise it (Burgess-Allen & Owen-Smith, 2010; Crowe & Sheppard, 2011).

Logic maps explore the logical connection between defined elements and their corresponding merits. Generally, logic maps encompass decision maps, dialogue maps, and argument maps. Logic maps are especially useful in situations requiring decisions or appraisal.

Another kind of map is decision map which explore decision options and the thinking behind these options. A decision map consists of actors, decisions, questions, options and sub-options, pros and cons, arguments and evidence – and their relationships. The map helps uncover the behaviour of an actor or decision maker. It is usually illustrated in a systematic, disciplined way (Francois, Blackwood, & Jowitt, 2002; Kingsford & Salzberg, 2008). The process of decision mapping includes 1) identifying the problem, 2) exploring the options, 3) evaluating the options, 4) building solid foundations, 5) taking a critical view of the options, 6) finding the assumptions, and 7) weigh everything to find the best solution (van Gelder, 2009). Decision maps have the potential to minimise the biases of the author, and it aids in providing a shared understanding among decision players. Their only shortcoming is their scope is limited to decisions only.

Dialogue mapping when applied to recorded conversations, produces a collective reflection on complex problems. It has three elements of 1) shared a display, 2) active listening, and 3) argumentation scheme (Conklin, 2005, p. xii). Dialogue mapping as a technique challenges the traditional process of linear thinking in problem-solving, i.e. that solution comes from an issue in an orderly way (Conklin, 2005, p. 8). Also named as 'Argumentation map', dialogue maps have the potential to show alternative scenarios and different viewpoints (Horn, 2001).

Issue mapping or issue-based information system (IBIS) is a technique for dialogue mapping. IBIS consists of the following three elements: 1) Issues (or questions), 2) Ideas (or positions), 3) Arguments: Pros (arguments for) or Cons (arguments against) an issue (Burgess Yakemovic & Conklin, 1990). Issue mapping has a potential to uncover a wicked problem as in project management issues.

Causal maps are a special type of flow map that conveys the flow of causality. Causality and correlation in a system can assist a researcher in modelling and predict system behaviour. Causal maps uncover the causal relation among concepts and variables in a system. Application of flow maps in project management is numerous including and not limited to planning, governance, management and research. Although causality or correlation are investigated in research, mapping the relationship helps to evaluate and improve findings through a constructive process of polishing and refinement. Hence, causality maps will enrich any qualitative strategy of research by offering a big picture of the ultimate relationship in the collected data for further validation.

An integrated representation of assorted visual maps that individually and severally support each other to convey a message may consist of a number of illustrative

Figure 6.

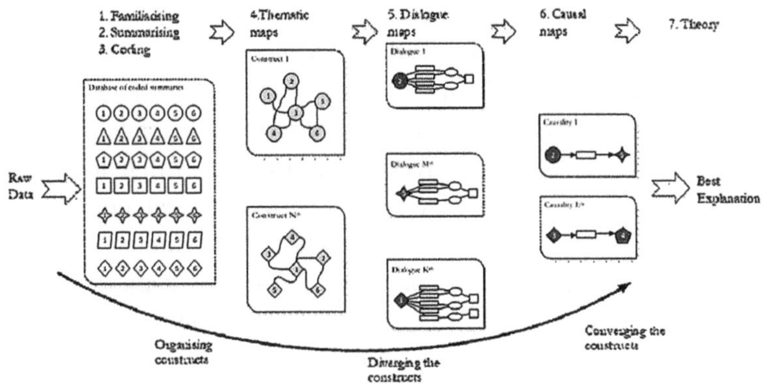

models or semi-structured complex model. An example is shown in Figure 6 (Zarei 2018). Maps may be used or combined in in one page such as posters, info-graphs, or murals to present ideas knowledge. However, they can only give an overview and normally do not cover the depth of content.

Posters and One Pagers

Posters combine many of the visual tools described earlier and are used to organise complex information and to present them in a succinct manner. A poster is a useful way of communicating only the essential facts in a project to the reader. Posters are usually combined with a short presentation. Posters and poster presentations are widely used in business and industry conferences as well as in higher education (e.g. Deonandran et al. 2013).

Posters present many advantages over traditional oral presentations. The reader can linger over a poster for as long as he or she wishes, and information is presented in a concise manner. Posters communicates scientific work in a highly visual way and enables graphic flexibility. It also enables a dialogue with viewers and creates opportunities for comments, questions, clarifications and mutual interaction. Posters are used widely for research presentations and academic exercises. Many universities print their own guide for students (e.g. University of Wisconsin-Madison, 2018; University of Melbourne, 2018; Asper, 2010; Hess et al., 2011; Radel, 1999), although these tend to be generic one size fit all- concise guides.

Gosling (1999) notes that there will be situations where posters have disadvantages in cases where the research have large volumes of mathematical analysis or complex experiment or where it is difficult to transport physically. In particular, Briscoe (1996) wrote about the misuse of poster presenting them as article presentations.

The layout and graphic design of the poster is extremely important in attracting and holding the interest of the reader. The reader is likely to look at the heading then at the conclusions and only after satisfied that the topic is of interest will the reader proceed to look at the contents. Therefore, the contents and layout of posters should reflect this, (Reybolds & Simmons, 1984). In industry and academic conferences, posters are used in settings where the reader is confronted with many other posters. Therefore, posters should stimulate interest rather than present complex details.

Similarly, one-pagers are like mini posters where they summarise only essential information on a single page such as a worksheet. Worksheets use a sequence of visual tools to help users see visual solutions in a logical and sequential manner. Perhaps a good example would be the problem-solving worksheet or the A3 worksheets first used widely at Toyota Corporation. In the A3 worksheet, a sequence of activities using other visual tools such as the Pareto charts and Fishbone diagrams to guide the user into arriving at the root cause of a problem. Correctly determining the root cause of the problem is crucial to finally solving the problem.

SHOWCASE: USE OF POSTER EXERCISE IN POSTGRADUATE ENGINEERING TO ORGANISE ENGINEERING INFORMATION

In this section we showcase a poster exercise that was given to postgraduate students as part of the ongoing assessment to help students learn about complex project organisation.

Students in a Project Management class are given a case study of a typical large engineering project - in this case F-35 Joint Strike Fighter project. The class size was 192 students. The students were organised into groups of five or six people. They were then given detailed information where to find major reports relating to the project. Some of the important project documents uploaded to the University's Learning Management System (LMS) but it was expected that students do their own further research to obtain the relevant information. The groups were then allocated a topic relating to specific concepts in Project Management. There were a total five topics and each of these five topics dealt with a practical aspect of project management such as roles, systems management and procurement strategy, systems engineering and project initiation. The students had to work in groups outside of the class time to develop a poster and use some tutorials for discussions. They also have to make a ten-minute presentation using the poster to the rest of the class by the end of the semester. The students were told that the learning objectives of this exercise was the ability to analyse a selected topic in a real project management situation and present the relevant information in an engaging manner which will facilitate meaningful discussions. The poster and the presentation were graded exercises which made up

Table 2.

Criteria	Details
1. Content	Scope Coverage
	Insight
	Design of Poster
2. Presentation	Explanation
	Use of the poster
	Style
	Timing
	Ability to answer questions

10% (7% for poster and 3% for presentation) of the final marks for the subject. In the poster exercise, the students were given a list of project documents for pre-reading at their own time. During one of the subsequent tutorials in class, they were then given time to discuss the topic, given guidance on how to organise the information and to present it in an engaging manner. A lot of these reports relating to the project were in the public domain and students were expected to find relevant information on their own.

Figure 7.

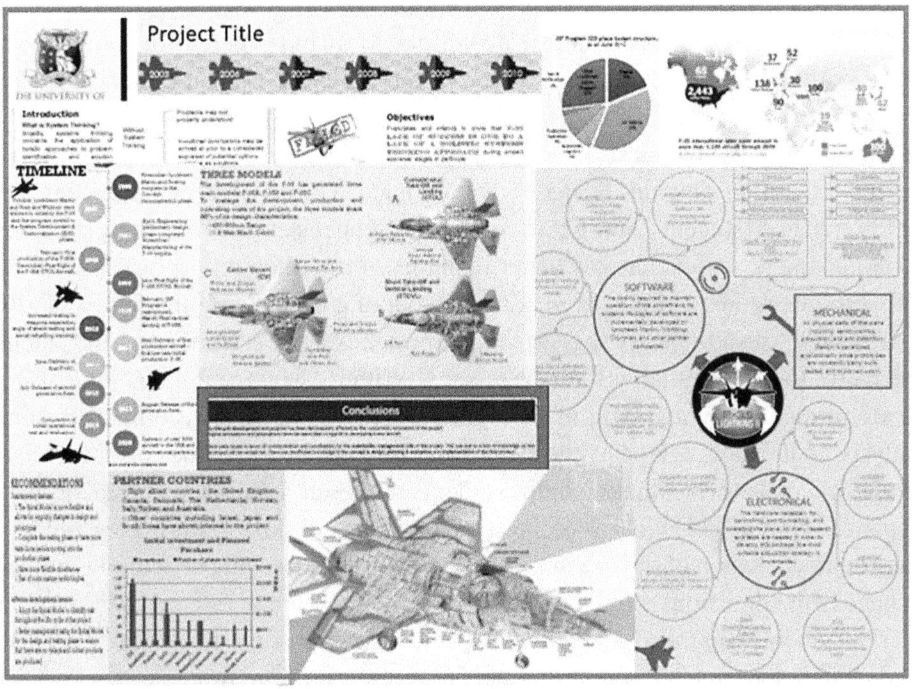

At the completion of the exercise, the students were asked to reflect on their experiences in the poster design activities and presentation. A survey tool mapping on learning using Bloom's taxonomy was administered. The data was collected during class time using a secure online tool. The posters were marked by a panel of six judges based on the following criteria (Reimann n.d.):

DISCUSSION

Research on Use of Poster in Engineering Education

The reflection that the students undertook was used as part of our research into the use of posters to see how it had helped learning and if it indeed enhances learning (Hui, Zarei and Duffield 2015). Similar work was done at the Vocational education and Training (VET) level (Hui & Zarei, 2016). Students reported that they found that visual tools such as a poster brings about more effective learning than text-based exercises.

Of the 192 students in the class, 65 students responded and provided complete answers to the survey tool. When asked to compare with a written assignment, 57% of the students agreed that posters positively help them relate to real-world project situations and 57% reported that it had positively helped them think about their own behaviour as project managers and 58% reported that it positively enhanced their knowledge of the behaviour of a Project Manager. More importantly, the respondents reported the greatest educational effect in the affective domain, confidence in dealing with others (71%), reading abilities (60%), listening skills (54%) and giving and responding to feedback (55%). The educational effects were marginal in the cognitive learning domain. Posters when coupled with the presentation exercise clearly enhanced students' abilities to communicate succinctly (79%). Many students also rated the poster exercise as better than that of a written report (refer Table 3) with attribute scores ranging from 3.2 to 3.6 out of a possible score of 5.

The survey results clearly suggest that the poster exercise can be strong and viable alternative to a traditional written exercise. A written assignment helps students to develop critical analytical skills in the cognitive areas of knowledge, comprehension, analysis, application, synthesis and evaluation (Krathwohl 2002), the poster exercise involving a real-world situation helps students to present issues and arguments and help in developing behavioural skills on top of analytical skills. This makes a strong case for the inclusion of the use of poster design in engineering curriculum. project management education.

The students were also asked about specific aspects of the poster exercise that needs improvement. It is noted that students would like more time for presentation time so that they can get practice their presentation skills. It is also noted that students found the interactive discussions of the presentations very useful. In particular when they had to answer questions posed by the audience after the presentations. There were also comments that sine groups felt that they benefitted from direct guidance from tutors during the practice classes.

Poster as an Experiential Learning Tool

Poster is a form of experiential learning that can be mapped to Bloom's taxonomy (Krathwohl 2002). Examples of skillsets in the cognitive domain and the affective domain include receiving information, giving information and examples of behavioural domain skillsets include working collaboratively, responding to questions and communicating succinctly. Through active learning (Freemana et al. 2014), poster exercises help to stimulate thinking by forcing students to make judgements about what are the important key issues in real life situations. It was noteworthy that the benefits in terms of cognitive skills developed across all affective domains and enhancements to presentation skills, communication are related to the expected behaviour of professional project managers.

Educational Implications and Managerial Significance

In this section we will explain how students can learn using visual learning tools in a constructivist classroom. In the constructivist classroom, students create their own knowledge by building interactively on what they know by creating solutions to problems. The emphasis of the learning process, therefore, shifts from the teacher distributing knowledge to students to a system where the teacher facilitates learning through group work or individual projects. (Bada 2015) The learners gain ideas in an increasingly complex manner where they build solutions and thing, get feedback, reflect and learn (Bargard, Corsaro & Ewaney 2014).

Another theory which explains action research type is inquiry–based learning where students where students are required to engage complex reasoning skills to solve problems especially at higher education level (Briggs and Tang 2007). This contrasts with a task-based lab where students are given steps to perform an experiment in a "cookbook" manner with a set sequence and recipe to arrive at a desired outcome. In relation to the problems faced by project management educators, it will be interesting to review and apply some of theories of how people learn to the area of project management. Bloom's taxonomy (Krathwohl 2002) provides a theoretical framework for how people learn. The first learning dimension or learning domain is the cognitive domain which categories learning into remembering, comprehension, application, analysis, evaluation and creation. These cognitive processes can be applied to four knowledge dimensions of factual knowledge (basic facts), conceptual knowledge (interrelationships between objects in a larger structure or system), procedural knowledge (knowledge of how to do something) and metacognitive knowledge. (self-awareness and knowledge of one's own knowledge). All of these four knowledge dimensions are widely found in the field of project management. The second domain, the affective domain (Krathwohl, Bloom & Masia 1964), deals with receiving, responding, valuing, organising and internalising. In this domain, the learner appreciates, feels enthusiasm or motivation for learning. In a project management environment, this translates into abilities to value opinions, have feelings of engagement or enthusiasm and internalising knowledge. The final learning domain is the behavioral or psychomotor domain (Simpson 1966) which deals with physical movement and manual tasks. It includes how people perceives non-verbal cues, responses to different situations, proficiency in movement or habit formation and adaptation based on existing skills. In the project management, this translates into a project manager's behavior under different situations.

The educational aspect of developing the skillsets in the three learning domains have significant managerial implications as these are the essential communication skills that engineers are expected to possess. Cognitive skills such as the ability to quickly comprehend information and organize data for analysis are important skills for critical analysis and problem-solving. The affective domain skillsets involve dealing with people and stakeholders on the job are basically soft skills that translates into ability to deal effectively with people using technical information. Finally, the psychomotor domain are important as these the skillsets that influence and engineer's behavior.

CONCLUSION

Visual tools used in presenting data and information are important aspects of engineering data visualisation. As such they should be given more emphasis and included in engineering graphics curriculum. In this chapter we reviewed major visual tools such as the graphs, charts and posters. We also showcased an example of the use of a poster exercise to help students learn how to visualise data and information. Our study indicated that the students found the poster to be beneficial. When compared to a conventional written assignment, the poster exercise offered advantages in cognitive skills development.

Although engineering information has traditionally been communicated using engineering drawings and engineering graphics, engineers can also use visual tools such as graphs and diagrams generated through computer generated graphics to visualise data. Visualisation of product and processes, knowledge of visual tools such as graphs, charts, fishbone diagrams, and one-pagers, posters etc are important as they have been successfully employed in the industry for decades as a basis for organisational learning and dissemination of information. It is an important part of an engineer's tool kit and skill sets. In building such skillsets, use of text-based learning can be complemented by visual tools. As shown in some of research, visual learning tools overcome some of the disadvantages of language by showing complex relationships in diagrams or in a visual manner. This has educational implications in that a proper structured poster exercise can help learners acquire skills, knowledge and know-how. The proper use of visual tools also has significant managerial implications as visual tools can be used to represent ideas in a concise, logical and consistent manner.

REFERENCES

Ackoff, R. L. (1974). *Redesigning the Future: Systems Approach to Societal Problems*. Wiley.

Asper, N. (2010). *Instructions for poster presentations*. Retrieved 1 June, 2011, from http://www.tcnj.edu/~asper/posterrules.html

Bada, S.O. (2015). Constructivism Learning Theory: A Paradigm for Teaching and Learning. *IOSR Journal of Research & Method in Education, 5*(6), 66-70.

Bargard, A., Corsaro, M., & Ewaney, M. J. (2014). The Poster Presentation as Experiential Education: Bridging the Gap between Research and Practice. *Clinical. The School Review, 7*(2), 105–108. doi:10.1891/1939-2095.7.2.105

Briggs, J., & Tang, C. (2007). *Teaching for Quality Learning at University*. New York: McGraw Hill.

Briscoe. (1996). *Preparing Scientific Information*. Springer-Verlag.

Burgess-Allen, J., & Owen-Smith, V. (2010). Using mind mapping techniques for rapid qualitative data analysis in public participation processes. *Health Expectations : An International Journal of Public Participation in Health Care and Health Policy, 13*(4), 406–415. doi:10.1111/j.1369-7625.2010.00594.x PMID:20550595

Burgess Yakemovic, K. C., & Conklin, E. J. (1990). Report on a development project use of an issue-based information system. In *Proceedings of the 1990 ACM conference on Computer-supported cooperative work - CSCW '90* (pp. 105–118). New York: ACM Press. 10.1145/99332.99347

CESE. (n.d.). *Developing a logic model.* Retrieved from https://education.nsw.gov.au/teaching-and-learning/professional-learning/evaluation-resource-hub/media/documents/Develop-a-logic-model-presentation_v2.pdf

Churchman, C. W. (1967). Wicked Problems. *Management Science*, 141–146. doi:10.1287/mnsc.14.4.B141

Conklin, J. (2005). *Dialogue Mapping: Building Shared Understanding*. Wiley.

Crowe, M., & Sheppard, L. (2011). Mind mapping research methods. *Quality & Quantity*, *46*(5), 1493–1504. doi:10.100711135-011-9463-8

Deonandran, R., Gomes, J., Lavigne, E., & Blanchard, R. (2013). A pilot study: research poster presentations as an education tool for undergraduate epidemiology students. *Advances in Medical Education and Practice, 4*, 183-188.

Francois, J.-C. B. A., Blackwood, D. J., & Jowitt, P. W. (2002). Decision mapping : Understanding decision making processes. *Civil Engineering and Environmental Systems*, *19*(3), 187–207. doi:10.1080/10286600214151

Freemana, S., Eddya, S. L., McDonougha, M., Smith, M. K., Okoroafora, M., Jordta, J., & Wenderotha, M. (2014). Active learning increases student performance in science, engineering, and mathematics. PNAS, Psychological and Cognitive Sciences, 111(23).

Funnell, S. C., & Rogers, P. J. (2011). *Purposeful Program Theory:Effective Use of Theories of Change and Logic Models*. Hoboken, NJ: Wiley.

Gosling, P. J. (1999). *Scientist's guide to Poster Presentations*. New York: Springer LLC. doi:10.1007/978-1-4615-4761-7

Government, N. S. W. (2016). *NSW Government Program Evaluation Guidelines*. Sydney: NSW Department of Premier and Cabinet.

Hess, G., Tosney, K., & Liegel, L. (2011). *Creating Effective Poster Presentations: An Effective Poster*. Retrieved 1 June, 2011, from http://www.ncsu.edu/project/posters

Horn, R. E. (2001). Knowledge Mapping for Complex Social Messes. *Knowledge Creation Diffusion Utilization*, 1–12. Retrieved from http://www.stanford.edu/~rhorn/SpchPackard.html

Hui, K. P., & Zarei, H. (2016). Using inquiry-based visual tools to aid inclusiveness in vocational education. *The 25th National Vocational Education and Training Research (NCVER) Conference*.

Hui, K. P., Zarei, H., & Duffield, C. (2015). How do we instil project management experience into young engineers? The use of posters as a learning tool. *Proceedings of the Australian Association for Engineering Education 2015*.

Jackson, K. M., & Trochim, W. M. K. (2002). Concept Mapping as an Alternative Approach for the Analysis of Open-Ended Survey Responses. *Organizational Research Methods*, *5*(4), 307–336. doi:10.1177/109442802237114

Juran, J. M. (Ed.). (1974). Quality Control Handbook (3rd ed.). McGraw-Hill Book Company.

Kingsford, C., & Salzberg, S. L. (2008). What are decision trees? *Nature Biotechnology*, *26*(9), 1011–1013. doi:10.1038/nbt0908-1011 PMID:18779814

Krathwohl, D. R. (2002). A Revision of Bloom's Taxonomy: An Overview. Theory Into Practice, 41(4).

Krathwohl, D. R., Bloom, B. S., & Masia, B. B. (1964). *Taxonomy of educational objectives: The classification of educational goals. In Handbook II: The affective domain*. New York: David McKay.

Lieu, D. K., & Sorby, S. (2009). *Visualization, Modelling and Graphics for Engineering Design* (1st ed.). Cengage Learning.

Mastropieri, M., & Scruggs, T. (2004). The Inclusive Classroom: Strategies for The Inclusive Classroom: Strategies for Effective Instruction Effective Instruction (2nd ed.). Pearson.

Radel, J. (1999). *Creating Effective Posters*. Retrieved 1 June, 2011, from www.kumc.edu/SAH/OTEd/jradel/Poster_Presentations/PstrStart.html

Reimann, N. (n.d.). *Getting to grips using posters for formative feedback*. Northumbria University. Retrieved from http://www.northumbria.ac.uk/static/5007/cetlpdf/signpost9.pdf

Renolds & Simmonds. (1984). *Presentation of data in Science, Publications slides, posters, overhead projections, tape=slides, television, Principles and practices for authors and teachers*. Dordrecht, The Netherlands: Marinus Nijjhoff Publishers.

Rittel, H. W. J., & Webber, M. M. (1972). *Dilemmas in a General Theory of Planning*. Working Paper No. 72 – 194. University of California, Berkeley.

Simon, H. A. (1984). The Structure of Ill-structured Problems. In N. Cross (Ed.), *Developments inDesign Methodology* (pp. 317–327). Chichester, UK: J. Wiley & Sons.

Simpson, B. J. (1966). The classification of educational objectives: Psychomotor domain. *Illinois Journal of Home Economics*, *10*(4), 110–144.

University of Melbourne. (2018). *Academic Posters, All about academic posters.* Retrieved 03/07/2018 from https://services.unimelb.edu.au/__data/assets/pdf_file/0007/470059/Academic_posters_Update_051112.pdf

University of Wisconsin. (n.d.). *The Writing Cneter Academic and Professional Writing: Creating Poster Presentations.* Retrieved from https://writing.wisc.edu/Handbook/PosterPresentations.html

van Gelder, T. (2009). *What is Decision Mapping?* Retrieved February 1, 2017, from https://timvangelder.com/2009/09/08/what_is_decision_mapping/

Zarei, H. (2018). *Participants' power asymmetry in public infrastructure projects* (PhD thesis). University of Melbourne.

Chapter 11
Strategies for Effective Teaching in Project Management

Ehsan Gharaie
RMIT University, Australia

Dallas Wingrove
RMIT University, Australia

ABSTRACT

Preparing project management students for their future life and work requires actively developing and evidencing a suite of transferable skills and attributes. This chapter reports on how a student-centered pedagogy, which included the use of guided sequential exercises, and the collection of instantaneous student responses through a personal response system (PRS), was implemented in a large first year undergraduate Project Management course. The students' perceptions of this pedagogy demonstrate that they found the pedagogical approach supported their learning and fostered deeper engagement in the course, with the most useful aspect of the course perceived to be its interactive nature. The chapter affirms the importance of giving life to a learning orientation conception of learning. The chapter has implications for ensuring learner engagement in the particular discipline of project management and for good practice in large class context in higher education.

DOI: 10.4018/978-1-5225-8452-0.ch011

PROJECT MANAGEMENT EDUCATION

Preparing Project management students for their future life and work requires actively developing and evidencing a suite of transferable skills and attributes. These attributes include professional communication, critical thinking, collaborative problem solving, and critical reflective skills. For project management graduates to transition to work, they must bring to the profession the capacity to integrate and apply their discipline knowledge and expertise in response to the diverse and competing demands which define the profession and diversity of contexts in which project management can be undertaken. As Pant and Baroudi (2008) identify, similar to many other professions, the role of the Project Manager comprises a complex hybrid of skills sets, with 'interpersonal ability, technical competencies, and cognitive aptitude' paramount (p. 124).

The demand for project management graduates and project management education is growing. The number of enrollments is increasing and there are undergraduate and postgraduate courses offered at established universities with well-known reputations and brands. There is also a demand from other majors such as engineering and business in which students are encouraged to undertake courses in project management as their elective. The increasing number of enrollments has led to larger class sizes.

In higher education, the large classes are often delivered in lecture mode. As the literature attests, the large class lecture can present many challenges for educators, with the risk that students engage in a didactic learning experience, whereby a one-way transmission of knowledge negates opportunities for them to actively engage in learning. There is well established evidence that one-way communication in the lecture setting does not result in good learning outcomes (Gysbers et al., 2011). A purely didactic approach, not balanced with student centred learning, can result in students retain only a small percentage of the learning material. Thus, the larger class sizes and the need for student engagement aiming for skill developments are two forces working in opposite directions.

The academics teaching project management courses often are not equipped with the right skills and knowledge to foster student engagement in large lecture settings. They may be highly experienced project managers who ran workshops and training courses in the industry. Or they may come from disciplines such as engineering, architecture, and construction management with less experience in dealing with large cohorts.

In the large lecture setting, technologies can help to foster a collaborative learning environment. These technologies can make real time formative feedback possible and create an engaging environment for learners. One of these technologies is Personal Response System (PRS) also known as Clickers or Electronic Voting System (EVS) (Surentheran, 2014). PRS has primarily been adopted to enhance the learning experience. It has been used in different contexts and subject areas (Liu and Taylor, 2013, Masikunas et al., 2007, Moss and Crowley, 2011, Voelkel and Bennett, 2014, Wolter et al., 2011).

The Personal Response System used in this course involved the use of Turningpoint and Responseware. By using PRS the students were able to contribute live their responses to the questions posed in class. This approach facilitated a learning process where the teacher and all students could immediately see the responses of the entire class. This was critical to promoting active learning as students could share their responses with each other, view those of their peers and receive just in time feedback. Using the PRS also provided an important mechanism by which the teacher could diagnose any aspects of student's learning that warranted targeted attention. By using the PRS student's active participation in the learning process was fostered. Whether and how the use of these technologies and the deployment of sequential learning activities foster student centered learning are explained in the this chapter. Whether and how the use of these technologies and the deployment of sequential learning activities foster student centered learning are explained in the this chapter.

This chapter provides a case study of a project management course. It examines how student centered learning and student engagement can be achieved in large cohorts of 150 to 300 students.

INTRODUCTION TO THE COURSE

The Project Management Techniques course is a foundation course that is more focused on the technical aspects of project management. In this course students learn how to apply knowledge of complex project management theories in their professional practice. The skills that they develop in this course include being able to apply project selection methods, to develop project plans using different techniques, and to collaborate with a team to develop their plans. The students conclude the course by developing a project plan that demonstrates their competencies in applying the project planning techniques.

The Project Management Techniques course is designed to enable students to:

- Apply selection measures to evaluate projects
- Select and apply appropriate project management techniques for effective project planning
- Collaborate with a team to develop a project plan for project stakeholders

There are major challenges in the delivery of this course.

Firstly, this course is a common first year course delivered across three different undergraduate programs. The student cohort is large with numbers reaching 300 in some semesters. The cohort is diverse and cross disciplinary and as a first year course, the course must also support student's transition into higher education. Such transition includes adapting to the large class context, which marks a significant departure from students' secondary education where they learn in classes of up to 25 students.

Secondly, project management techniques involve step-by-step processes. In order to effectively teach these techniques, students must engage in the process of working through each step. If a step is missed, the subsequent steps become incomprehensible.

For example, one such method is Critical Path Methods. Critical Path Method is a project planning technique that produces a time schedule for projects. The three steps involved in undertaking this technique are:

1. Drawing project activity network
2. Forward pass: calculating early start and finish dates of project activities
3. Backward pass: Calculating late start and finish dates of project activities

A project manager cannot undertake the third step without going through the previous steps. Similarly, students cannot practice and learn the third step without learning and practicing the previous steps. Therefore, it is imperative that students replicate and practice these sequential steps in order to effectively learn and successfully apply project planning techniques. Yet whilst necessary, the requirement for students to practice each step can work against student engagement. This is particularly challenging when teaching in the large class to a transitioning first year cohort.

Thirdly, the course includes a focus on Project Management techniques that are numerical which is highly challenging since the student cohort is generally not proficient in their knowledge and use of mathematical problem solving. It is critical that the requisite techniques are clearly unpacked and modeled with student's learning scaffolded throughout the course. It is also critical that students are actively engaged in the process of applying their developing knowledge and skills. It is fundamental that students receive formative feedback to support their learning.

Fourthly, this is a course covering fundamentals of project management techniques. In industry, these techniques are performed using software packages such as Microsoft Projects and Primavera. Students are aware of this and keen to use software. Part of the challenge in teaching this course is in being able to justify the importance of the fundamentals. Students understand this importance when they engage with the techniques and apply them. By learning the subtle aspects and intricacies of the techniques, they appreciate the value in deep understanding of project planning versus the ability to produce a project plan using software without understanding the processes performed behind the scene.

For all these reasons, student engagement in their learning is salient in successful teaching of the course. it is essential that learning is embedded in a cycle of knowledge acquisition with opportunities for students to apply their learning in practice. Passively listening to theory would not foster the course's intended learning outcomes nor ensure that the opportunities for students to be proficient in their use of the PM techniques are optimized.

The pedagogical approach adopted in the course was designed to actively engage students with the course content, providing formative learning activities in which students can apply the techniques learnt. A mix of student centred and teacher guided learning was implemented to ensure that students were both guided and actively engaged in learning by doing. The next section discusses the implementation of the teaching strategies in this course.

OVERVIEW OF THE COURSE DELIVERY

The course is delivered through weekly lectures and tutorials. Each week is dedicated to a particular Project Management technique such as Project Selection Methods, Bar Chart, Critical Path Method (CPM), Program Evaluation and Review Technique (PERT), cash flow management, and Earned Value analysis. However, the learning process goes beyond lecture and tutorial. Figure 1 demonstrates the overall delivery of the course. The following sections provide details for each part of the delivery and describe two examples of the delivery.

Table 1.

Assessments	Week 1	Week 2	Week 3	Week …
	Lecture • Step by step problem solving • Real time feedback	Lecture • Step by step problem solving • Real time feedback	Lecture • Step by step problem solving • Real time feedback	Lecture • Step by step problem solving • Real time feedback
Individual weekly formative assessments	Tutorial • Exercises • Problem solving method of learning	Tutorial • Exercises • Problem solving method of learning	Tutorial • Exercises • Problem solving method of learning	Tutorial • Exercises • Problem solving method of learning
Group project summative assessment	Development of project plan			

LECTURE

There are two key strategies to create an interactive lecture (Meltzer & Manivannan 2002). The first strategy is to provide deliberate step-by-step guidance for students to think about, discuss, and respond to a carefully designed sequence of questions and exercises. The second strategy is the provision of a system that collects instantaneous responses from students in class simultaneously. Both strategies have been implemented in this course.

STEP BY STEP PROCESS

Each lecture is divided to smaller sections that correspond with each step of undertaking a PM technique. The lecturer explains each step and demonstrates a practical example of that step, then students are asked to practice that step on a specific problem given to them. Students are given a limited amount of time to work on that section and reach the solution.

ONLINE POLL

Collection of instantaneous responses is the second strategy that is implemented through an online polling system. The online polling system used in this course is developed in TurningPoint software (www.turningtechnologies.com), and responses are collected through ResposeWare (www.rwpoll.com), which is an online platform. This platform is accessible via any device that can connect to the Internet. The use of ResponsWare is extremely successful because of the familiarity of students with their devices, and using their mobile devices satisfies their craving for checking their mobiles. The results of each poll are shown to the students in real time in graphs.

Allowing students to see their peer's responses and the variations in those responses fosters a learning environment whereby students are motivated to debate and contest the different choices their peers have made. It also shows students there are other students who think similarly to them.

In addition to providing real time feedback to students, the polling system enables the lecturer to adjust the class in accordance with student's learning needs. The lecturer also benefits from these polls. The results of the poll demonstrate the 'muddiest points' for the students and thus guide the lecturer as to where the learning focus needs to be. The poll supports the lecturer to efficiently manage the lecture time and breaks the monotonous lecture setting, making the learning environment more engaging. In addition, there is time available to the lecturer and students during the polls for one on one interaction.

Two examples of this practice are described in the following sections to clarify how the sequential exercises and the real time feedback are implemented. The examples contextualise the teaching method for teachers who would like to apply this teaching approach in teaching project management courses. They include two topics of Critical Path Methods and cashflow management.

EXAMPLE ONE: CRITICAL PATH METHOD (CPM)

As mentioned above, there are three steps involved in undertaking CPM:

1. Drawing project activity network
2. **Forward Pass**: Calculating early start and finish dates of project activities
3. **Backward Pass**: Calculating late start and finish dates of project activities

First students are shown how to undertake the first step of the process. The lecturer explains the process for drawing a project network and demonstrates this step through an example. Then, students are asked to draw the correct project network.

Figure 1 shows the question posed to the class. The students are given limited time during the lecture to undertake the activity.

Similar approach is taken with the step two of the process – forward pass. The lecturer explains how to undertake the step and demonstrate the implementation on the technique through an example. Then, students are asked to follow the same steps and practice their understanding by applying the technique on a question posed to the class. Students are asked to enter your answers into the Personal Response System.

Figure 1. Example of the first step in CPM undertaken by students

Step one: draw the network

Activity	Predecessor
A	-
B	A
C	A
D	A
E	B, C
F	B, C, D
G	F
H	E, G

Figure 2. The second step in CPM technique and the poll results

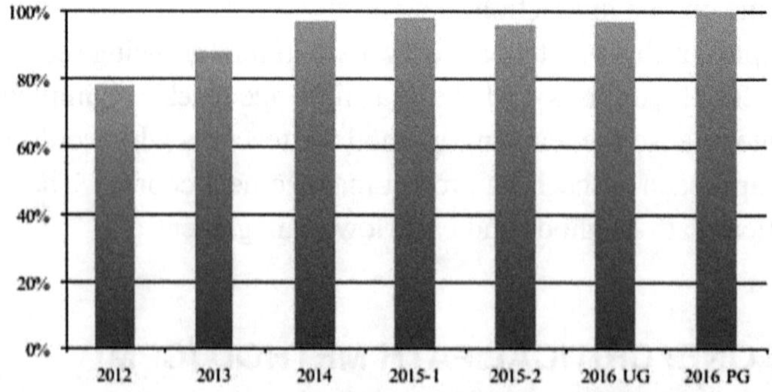

After the time given for the activity is exhausted, the lecturer closes the polling session and show the result of their answers to the class. Figure 2 demonstrates the implantation of the forward pass calculation and the real time poll.

Step three in the CPM technique is the backward pass. The following figure demonstrates the question posed to the students and the polling results.

Figure 2 and 3 show a sample of polling results. As can be seen in this figure, in this particular exercise, most of students could find out the correct answer. This is reassuring for the lecturer that the message got across and the class can move on to the next step.

Figure 3. The third step in CPM technique and the poll results

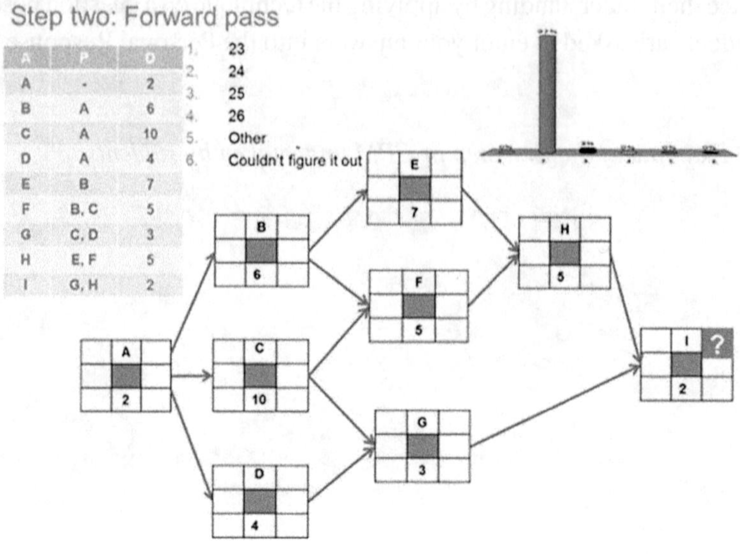

Figure 4. An example of polling with variety of answers

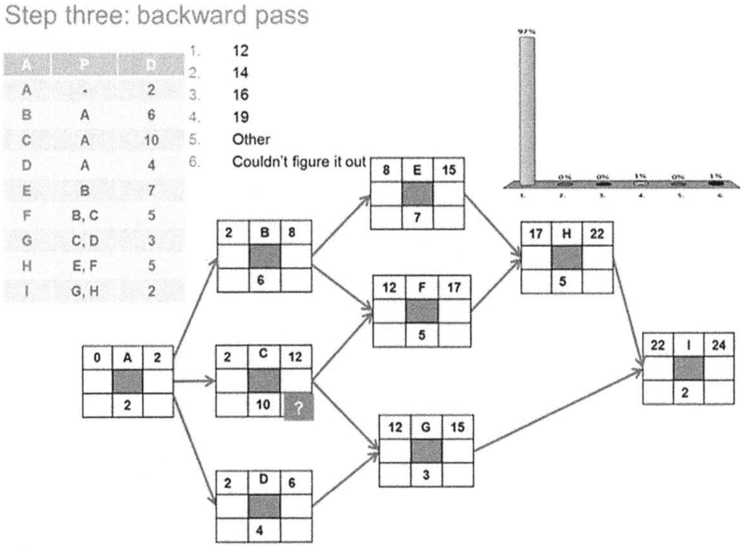

However, in some instances the polling results show a variety of answers. For example, Figure 4 demonstrates one of these instances. In this particular example, the correct answer is 12 (option 2). Captured responses, however, showed that answers varied between 12, 13, and 17. There were students who ended up with a different answer than these three options and have chosen 'other'. More importantly, 11% of respondents chose 'could not figure it out' which signifies that they did not understand the concept and they could not solve the problem at all.

EXAMPLE TWO: CASHFLOW MANAGEMENT

The second example is the cashflow management concept. The students need to learn how to forecast the outflow, inflow and netflow of the project. In each step they are introduced to the concept, they see a demonstration of the calculations, and they apply what they have learnt on a problem.

For this particular lecture the interactions of steps are as follows:

- Explanation of the overall concepts of outflow, inflow, and net flow in cashflow management
- First section- cash outflow
    - Outflow forecasting explained by the lecturer
    - Outflow forecasting undertaken by students

Figure 5. Example of project cost breakdown

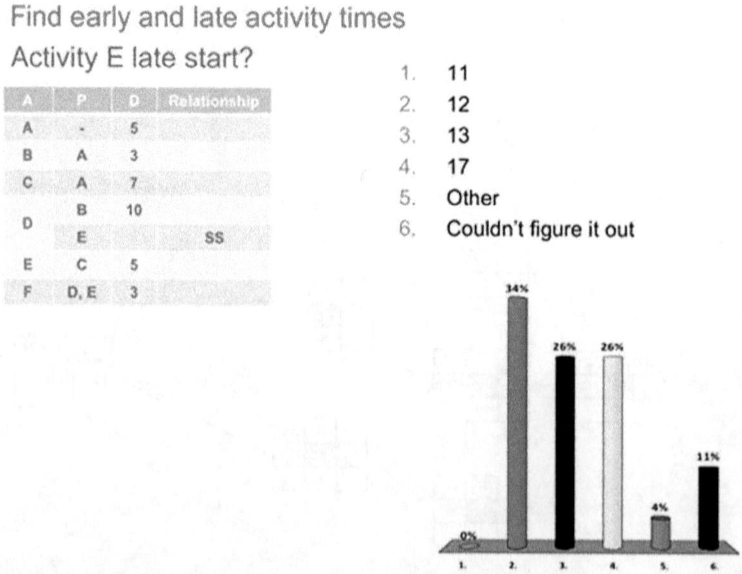

  - Response collection and instant feedback
- Second section- cash inflow
  - Inflow forecasting explained by the lecturer
  - Inflow forecasting undertaken by students
  - Response collection and instant feedback
- Third section- cash net flow
  - Netflow forecasting explained by the lecture
  - Netflow forecasting undertaken by students
  - Response collection and instant feedback

The first step of this process is explained in further details to provide concrete examples of how this method is applied in the lecture.

First the concept of cash outflow is explained. Students learn how project outflow can be forecasted and they see an example of such forecasts. Figure 5 shows an example of project cost breakdown which is the basis for outflow forecast.

Figure 6 demonstrates how the cost breakdown can be translated to an S-curve or cumulative outflow.

In the second step of the process, students have the opportunity of implementing what they have heard on a simple problem. This step reinforces their learning and give them the chance to find out if the sticking point in the process.

Figure 6. Example of calculation of cumulative outflow and S-curve

	Cost	1	2	3	4	5	6	7	8	9	10	11	12	13	14
A	10,000	2500	2500	2500	2500										
B	50,000			10000	10000	10000	10000	10000							
C	30,000				10000	10000	10000								
D	60,000								20000	20000	20000				
E	25,000											12500	12500		
F	5000													5000	
Monthly outflow		2500	2500	12500	22500	20000	20000	10000	20000	20000	20000	12500	12500	5000	

Figure 7. Question for students to undertake project cost breakdown

Cumulative outflow graph

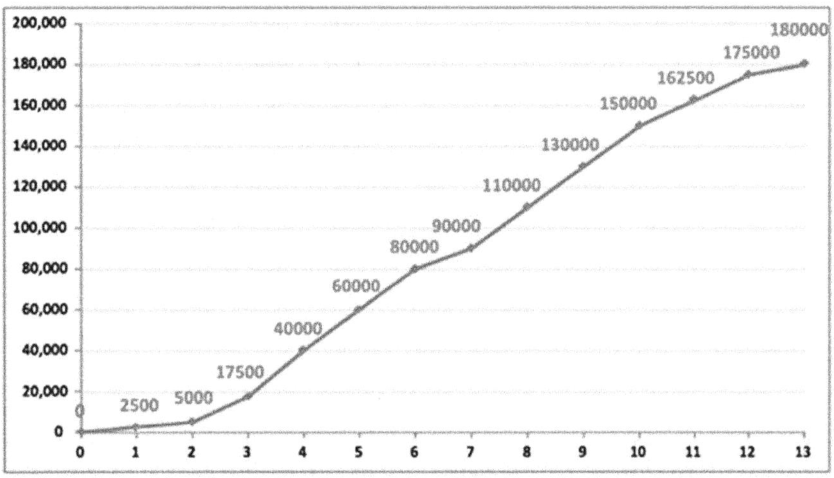

Monthly outflow	2,500	2,500	12,500	22,500	20,000	20,000	10,000	20,000	20,000	20,000	12,500	12,500	5,000
Cumulative outflow	2500	5000	17500	40000	60000	80000	90000	110000	130000	150000	162500	175000	180000

Figure 7 is the problem that students are asked to solve in the lecture right after they have heard about the concept of cash outflow and how it is calculated.

Students are given a limited time to solve the problem and enter their answer into a live online poll. The question related to this problem is shown in the following figure.

The responses are collected, and the solution is then explained for the class.

These steps are repeated for the second and the third section - cash inflow and net flow. By the end of the lecture not only students have heard about the concepts and

Figure 8. The question posed to the class related to Figure 7

	Duration	ES	Cost	1	2	3	4	5	6	7	8
A	2	0	$1,000								
B	3	0	$3,000								
C	2	1	$3,000								
D	1	3	$500								
Monthly outflow											
Cumulative outflow											

the way the cashflow parameters are forecasted, but they have implemented those forecasts and feel good about what they have learnt. The graphs and calculations that could be daunting become familiar concepts that the students feel confident in applying them.

The learning in the lecture is then reinforced in the tutorials through more complex and comprehensive examples. Because the students have already practiced their knowledge once, they are not afraid of the problems in the tutorial questions. They know they have already done it once and they should be able to do it again. That is why they are very engaged in the tutorial sessions although the tutorials are problem-based learning and there is no formal guidance provided.

The following section describes how the learning in the lecture is continued into the tutorials.

TUTORIAL

The knowledge and skills introduced to the students in the lectures is reinforced and developed in the tutorials. Tutorials are run as a problem-based learning exercises. These tutorials employ a different pedagogy to the lecture, where learning is more teacher guided. Conversely, the tutorials are relatively unstructured. There is a tutor available for students to answer their questions. The tutor, however, does not repeat the lecture. Students work individually or with their peers to apply planning techniques. The dynamics of the tutorials is fluid. Some students prefer to work individually and some work with groups. And, some students have expressed their interest in learning the concepts through teaching others. The flexibility and the problem-solving nature of the tutorials provides them with this opportunity.

In the tutorial sessions, students are asked to solve a few problems and apply their knowledge of project planning technique that they have learnt in the lecture. The problems in the tutorials start with simple problems and their complexity increases as students work through the problems. Every problem is a stepping stone to the next one. This ensures that learning is fostered incrementally. As students work through the questions, the level of discussion and questioning increases. The tutor joins in the group discussions and guides students through the sticking points. There is no need to explain how to solve the problem from start to end. Students have already determined how to approach the problem and there may only be a couple of points that they are debating. Their queries are supported by the tutor.

The tutorial problem-based exercises are formative in their design. Students submit their solutions on a weekly basis and receive feedback a few days before the next lecture. The tutorial exercises provide a means by which new knowledge can be consolidated and then applied in practice. The tutorial questions give the students the opportunity to consolidate their learning and clarify their 'muddy points'.

The tutorial sessions provide the opportunity for students to develop and extend their learning from the lecture. The students recognise the importance of the learning process designed and respond positively to this pedagogy.

GROUP PROJECT: DEVELOPMENT OF A PROJECT PLAN

An important part of the delivery of the course is the assessment task delivered in groups and aim at the development of a project plan. This is a summative assessment designed to consolidate and measure student's learning. Students are required to apply their project planning knowledge to a real case study and produce a project plan. This project plan includes a complete project scope, deliverables, exclusions, milestones, Work Breakdown Structure and activity definitions. The assessment is designed to enable students to apply and demonstrate the following project planning techniques:

- Project network
- Bar chart
- Critical path method
- PERT
- Crashing and reducing project duration
- Cash flow management

Students form groups at the beginning of the semester. They work within their group to complete the assessment task and deliver their project plan. The group work encourages them to collaborate on their project. Working collaboratively creates a learning experience where students share ideas and learn from one another. The focus on one project throughout the semester helps students to delve deeper in the planning issues of their project and understand this better. The group dynamics help them develop their teamwork and collaboration skills. The project also resembles the reality of work in the industry wherein strangers come together to form teams and deliver projects.

The planning techniques required in the assessment task directly are aligned with the techniques learnt each week during the semester. The alignment with the lecture contents and tutorial questions help students be ready for their project and feel comfortable approaching their project. It also consolidates what they have learnt in the lecture and tutorials and contextualises their learning in a more practical and real-world setting.

For example, this would be the learning process for students who choose to plan the construction of a residential building:

1. Learn how to use bar chart to plan a project in the lecture;
2. Practice the learning and skills on hypothetical questions in the tutorials;
3. Apply the knowledge on the planning of the construction project by drawing a bar chart for the residential construction project.

Applying their learning on a real project presents challenges for students which they have not necessarily anticipated. Every project presents its own complexities. These complexities encourage students to think more deeply about the planning techniques.

The learning in the course is compartmentalised based on the weekly topics. The group project, however, requires a single project plan that should be coherent and comprehensive. Undertaking the project shows student how different planning techniques are related and how different aspects of a project can be seen from different angles. The project supports students to develop a holistic understanding of planning techniques. Their developed knowledge and skills are in turn assessed summative.

EVIDENCE FOR THE SUCCESS OF THE TEACHING METHOD

The previous sections described how the sequential exercises were undertaken in the lecture and how the instantaneous responses were collected and communicated to the students. They also provided the context in which these learning activities are implemented and described how these learning activities are constructively aligned to the weekly tutorials and the final major assessment task. This section explains the results of this teaching approach by reporting the feedback received from the students through formal Course Experience Surveys. In this section, we discuss and analyse student's quantitative and qualitative feedback as captured via

Figure 9. The percentage of agreement with 'the teaching staff are extremely good in explaining things'

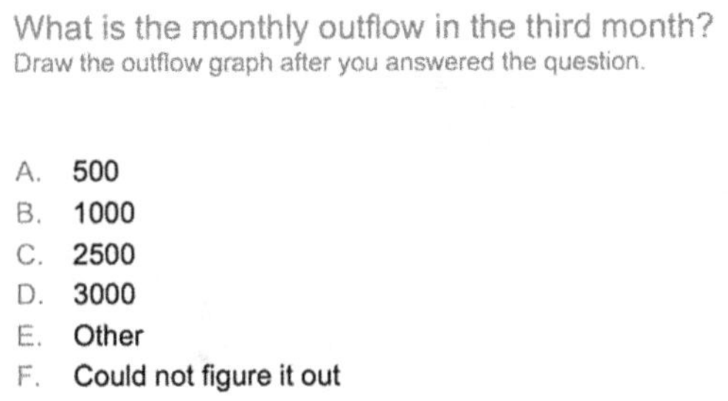

the Course Experience Surveys. These responses were captured over a five-year period, commencing in 2012.

These Course Experience Surveys are independently administered and are anonymous and voluntary. Student feedback is collected and reported via quantitative scoring and qualitative descriptive feedback. The authors report and analyse responses captured over five years from 2012 to 2016. The results of these surveys are analysed in relation to determining the effectiveness of the teaching method.

QUANTITATIVE FEEDBACK

The students' feedback in response to the statement 'the teaching staff are extremely good in explaining things' is shown in figure 3. The responses range from strongly disagree to strongly agree on a Likert scale. The graph reports the percentage of agreement with this statement.

This percentage has grown from 2012 and steadied since 2015 at 100%, which is the result of the introduction of sequential problem solving during the class. Between 2012 and 2015, the only significant change in the delivery of the course was the gradual introduction of the sequential problem-solving approach. At the middle of 2012 semester two, the sequential problems were introduced to the course. The collection of real time student responses was implemented in 2014. This percentage has grown from 2012. The most effective iteration was achieved in 2015 and it has not changed since then.

As can be seen in this figure, there is a very high percentage of students who perceive the course has been delivered efficiently and who recognise that lecturer effectively explained the course content. This outcome highlights the value of engaging students in sequential problem solving exercises and of ensuring that students are engaged in the course, with knowledge and skills development scaffolded appropriately.

Figure 10. The percentage of agreement with 'the teaching staff work hard to make this course interesting'

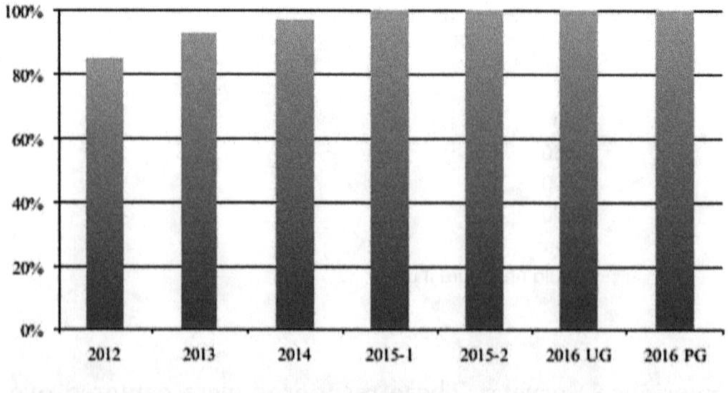

The use of the Personal Response System fostered student's engagement in the learning activities and course content. The next graph demonstrates the percentage of agreement with the statement 'the teaching staff work hard to make this course interesting'.

QUALITATIVE FEEDBACK

In this section, we discuss and analyse student's qualitative feedback as captured via the CES survey. In the survey, students are asked 'what was the best aspect of the course?' Students' answers focus mostly on one aspect of the course that they considered the best. Most answers were brief and to the point, however, about ten percent of students provided more detailed responses. Overall, 317 students provided descriptive qualitative feedback regarding the best aspect of the course. Their feedback was analysed to identify patterns and themes. We adopted an analytical approach, conducting comparisons across answers using the six-phase process outlined by Braun and Clarke (2006). This analytical approach entailed an iterative cycle of identification, analysis and review.

The data revealed that from the student perspective, the best aspect of the course related to interactivity deeper learning, employability and scaffolding learning. The following section discusses these themes and provides the examples of students' feedback.

Theme One: Fostering Interactivity

Interactivity is central to learner engagement. As the literature affirms, didactic learning approaches are not conducive to active participation in learning, and strike at the heart of the deep versus surface learning (Biggs 1999, Entwistle 1988, Ramsden 1992). As students responses demonstrate, interactivity was highly valued, with one hundred and eighty seven students (59%) identifying that actively engaging in the learning process was something they valued. This related to both the pedagogical approach in the lectures and the use of Responseware. Students' responses attest to the correlation between interactivity in the lectures and their engagement in the course.

Students identified how interactivity was facilitated in a number of ways including through the use of the polling tool Responseware. Students affirmed that they perceived that Responseware was a highly useful mechanism in providing them with instant and formative feedback. Such feedback was seen as an important contributor to the development of their disciplinary knowledge. Responseware was also viewed as highly beneficial to skill development, and to student engagement.

As Student A identifies

The use of the polling system is excellent and provides students with instant feedback on the course content

Further, students recognised and valued interactivity. By contrast to a didactic approach, students valued the mix of teacher guided and student centred learning, as Student B describes:

Having the opportunity to work through the techniques with the other students is an effective way to learn.

Students also identified the importance of ensuring opportunities for knowledge to be integrated and applied through the tutorial/workshop questions. They valued the links between the lecture and tutorial sessions.

Theme Two: The Teaching Approach

One hundred and ten (35%) of students identified that they highly valued the teaching approach adopted by the lecturer. In doing so they identified particular teacher attributes and pedagogy which they viewed as highly useful in fostering their learning and their engagement with the course content.

Students identified that they valued the teacher's focus on scaffolding their learning, explaining concepts thoroughly, and checking for understanding, as the following statements exemplify:

Student C:

The teacher clearly explains all material and allows you to attempt the work in the lecture and provides on the spot feedback.

Student D:

The teacher handles questions perfectly, he gives a clear answer and explains why something is wrong and assures you understand the reason why before moving on.

Student E:

Engaging subject matter delivered in a well sequenced and paced course.

Students also perceived that because they and their peers were engaged in the learning, there were very few behavioural issues which arose in class. As Student F describes:

Involving everyone and keeping everyone quiet at the same time made for a positive learning environment.

Providing opportunities for students to test and apply their knowledge was also perceived as critical to developing understanding.

Theme Three: Deeper Learning

Sixty-two respondents identified the role of the tutorial in supporting their learning. Student's feedback in relation to the tutorials highlighted that they perceived that the opportunities to apply their developing knowledge and skills in these sessions played a key role in their learning.

As Student G describes

The weekly tutorial problems were really enjoyable as we were able to apply the lecture material learnt.

Learning activities which build on concepts learnt in the lectures were also viewed as enabling understanding as the following comment by Student H highlights:

The weekly tutorial problems helped [...] me [...] understand all of the information we had learned in the lecture that week.

Students also valued the course design, with student feedback highlighting the importance of constructively aligned curriculum as the response here from Student I exemplifies:

What you learn in the lectures is what you do in tutorials and in assessments.

Students also identified that the structure of the tutorials, by providing opportunities to develop their knowledge and skills, also helped to foster their autonomy as learners:

Student J identifies such autonomy describing the tutorials as providing:

a platform for revising the entire teaching....I am proud to say that I can do the techniques without any hesitation.

Students also valued the opportunities to learn through interactions and problem solving with their peers. Whilst there are many issues which can arise for students when working in groups, opportunities to learn together were valued:

I like breaking up into groups and completing the tutorial work together. Other group members point out things to remember and making studying much easier. (Student K)

Theme Four: Employability

The integration of learning and work also emerged as a theme. Fifty-seven respondents referred to the relevance of the course content to real life professional practice contexts. Students identified explicit links between the course and their current or future professional practice which also fostered their enjoyment and interest in the subject material.

Students identified specific dimensions of learning, both knowledge and skills, which they valued. They perceived that their industry connected and relevant project management skills and techniques were actively developed through the course. Such identified skills included time management skills, with the taught Project Management techniques viewed as relevant to professional practice, practical and logical.

Students valued that the course design and learning activities enabled them to utilise a range of project management tools and methods applicable to the life cycle of a project. As Student L identifies, these methods were directly relevant to "the process of project planning and execution".

Students consistently reinforced that they valued "learning things that can be directly applied to the workplace' (Student M). This was the case for students who were both currently working within the Project Management field but also for those who intended to enter the profession:

Student N:

I can already tell its going to help me in the construction industry.

Student O:

The course allows you to understand and learn a variety of different techniques that will be useful later in the workforce.

Student P:

All techniques that I studied are very useful for my career. It is exact the same situation that I experienced in real work. This course helped me to apply these techniques to deal with my work effectively.

Beyond preparation for professional practice, students also perceived that their learning was relevant to their lives more broadly:

Student Q:

I feel the group project put everything learnt throughout the year into place and related it to real life.

Student R:

I found the content really relevant to real life situations.

Theme Five: Scaffolding Learning - The Constructively Aligned Curriculum

The final theme to emerge from the data relates to overall design of the course and the learning process. Thirty-one respondents explicitly mentioned this factor as the best aspect of the course. Students placed a high value on the clear links between developing their theoretical knowledge and applying their knowledge and skills.

Having a lecture at the start of the week to learn a new concept and then consolidating the teaching in a graded worksheet which you are able to complete in the tutorial with the assistance of your tutor made it extremely easy to learn all the concepts throughout the course. (Student S)

Clear, concise and well-structured course materials were also valued as the following comment indicates:

Course materials were also very organise, making it easy to refer back to material when needed. Slides were concise and very easy to understand; very helpful as this subject is unfamiliar to me. (Student T)

This comment further highlights the importance of scaffolding learning, particularly through an introductory course.

Again, making explicit links between the lectures and tutorials and assessed learning was reinforced by students as promoting learning:

Structure makes sense; we're taught the concepts in the lectures then we're encouraged to apply them in the tutorials. (Student U)

The importance students placed on learning being scaffolded is further revealed through these responses:

the teacher explains the techniques starting from the simplest aspects therefore it is easy to keep up and not feel overwhelmed and lost. (Student V)

The course has been set up incredibly well. You are engaged each and every week and taken progressively through the assessment requirements. (Student W)

THE PEDAGOGY

In light of sector wide low tertiary student survey response rates, student engagement in this course interestingly appears to have translated into consistently strong response rates over a five-year period. The average response rate was 32 per cent in this period and the number of responses has consistently been statistically sufficient. Such response rates include the provision of optional qualitative feedback. It is noteworthy that the students themselves elected to firstly complete the surveys and to secondly provide additional qualitative feedback which was highly positive, and which provides an interesting and most useful insight into their individual learning journeys. As the combined data highlights, students highly valued the opportunities to engage in active and student centred learning.

Tangey (2014) importantly identifies that the term student centered learning can be viewed as ubiquitous and is certainly a term that is ill defined in the literature. What is clear in the literature is that the term student centered learning is inextricably linked to constructivism. It is the term which recognises knowledge and the process of knowledge acquisition as a social construct.

Tangey (2014) also provides a useful way of conceptualising student centered learning and identifies two distinct yet interrelated aspects. Firstly, a learning orientated conception of learning, as distinct from didactic knowledge transmission, is one which adopts a holistic perspective of the student. This humanistic approach, whereby engagement with the student draws upon their wider context of learning, is also borne out in student's particular voluntary responses about their lecturer in the Project Management Techniques course. Our data affirms that students clearly valued the lecturer and that they perceived and experienced him as someone who cared about their overall learning and development, rather than perceiving their development as discrete, and bounded by the content of particular course. This is evidenced in a number of ways including how the course content related to student's future life and work, and how students perceived and experienced their lecturer as someone who encouraged, supported and motivated them, and through the assessment design, which was both developmental and realistic.

Secondly, Tangey's learner orientated conception of student-centered learning focuses on student's intellectual development, whereby students are viewed as "active participants rather than passive receivers" (p. 267). As students' responses illuminate, core to what students valued were the opportunities the teacher provided for them to apply their developing knowledge and skills, in the tutorials/workshops and through the completion of sequential exercises. Both elements were viewed as highly useful to enable learning. Students also awarded a high value to the use of polling tools as a way of tracking their own development and as a mechanism to support and enhance their learning through an interactive engagement with their teacher and peers.

Student's CES survey responses both qualitative and quantitative confirm the value of purposeful active engagement. Through a learning and teaching cycle whereby their knowledge was scaffolded, and whereby they were in turn provided with opportunities to integrate and apply their developing knowledge and skills, students perceived that they were supported to create their own understanding. They actively built upon prior knowledge to deepen their learning, whilst their autonomy through the learning process was encouraged and fostered. As the literature makes explicit this represents a critical element for effective learning outcomes (Kember & Leung, 2005a; Nicol & Macfarlane-Dick, 2006)

A learning orientated outcome was enabled in part by an underpinning coherent course structure, whereby student's prior knowledge was not assumed but rather enabled in part through sequential exercises and experiential tutorials. Importantly, students provided rich and positive feedback about the interactive teaching approach.

As the literature identifies, if students perceive teaching as 'good' they are more likely to adopt a deeper approach to their learning (Baeten et al., 2010). There is a body of research which identifies a link between student's perceptions of good teaching and a deeper approach to learning with a correlation between the two well documented in the literature (O'Neill & McMahon, 2005; Wilson & Fowler, 2005). The students who prefer deep learning also prefer cooperative learning (Baeten et al. 2016).

As discussed, the students also identified the teacher as an important contextual factor in their learning. The teacher's approach and presence contributed to student's positive experience of the course. This included class behavioural management. Students experienced the teaching as orientated towards them with disciplinary knowledge viewed as communicated with passion. The teacher's interest in the course content and in the students, themselves was highly valued.

Workload is also a factor that can influence students' approach and level of motivation and this was also identified by the students with work load viewed as realistic, rather than overwhelming. As Baeten et al. (2010) identify, there is a direct correlation between workload and deep or surface approaches to learning, with the former resulting from a measured and realistic approach. A body of research has shown that an excessive perceived or inappropriate workload is positively associated at a significant level with a surface approach (Kember et al., 2008; Kember*, 2004) and significantly negative with a deep approach to learning (Cope & Staehr, 2005). Kember* (2004) offers a possible explanation for these findings arguing that students resort to short cuts and a surface approach to cope with the perceived excessive assessment demands. The teaching approach importantly also facilitated on going feedback, including through the use of the PRS.

As Baeten et al (2010) identify, research evidences that there is a link between student's perceptions of their learning as relevant to their future practice and their adoption of a deeper learning approach. It is noteworthy that students identified a link between their learning and their profession (current or future) in Project Management.

In reporting the results from a study which investigated the efficacy of program design within a UK Master's program in Project Management Bentley et al. (2013) advocated for a research led evidence-based approach to curriculum design and delivery. We concur and argue for a pedagogy which enables the teacher in partnership with students to foster interactivity between teacher and students and between students. Such pedagogy is intended to foster learner engagement and create a learning culture where students and teacher are feeding back and forwards as they work within the course. This interactivity includes the provision of authentic learning experiences that ensures effective engagement through real life engaged learning which puts the student at the center of the learning experience.

CONCLUSION

This chapter has presented the results of the implementation of sequential class exercises and the use of Personal Response System in a large project management course. The student feedback captured over five years has demonstrated the effectiveness of these strategies and their usefulness when teaching a large cohort. As reported, students found the teaching method very effective and perceived and experienced the learning approach as developmental and engaging. We have demonstrated the importance of student-centered pedagogies in creating an environment in which active learning can take place. If we are to prepare our students for futures unknown, it is imperative that students are actively engaged in learning; that they are able to incrementally integrate and apply their developing knowledge and skills to ultimately prepare our students for their future professional practice.

This study has identified key elements that from the student perspective were conducive to fostering their learning: coherent logical course materials, interactivity through tutorials and lectures, which includes as facilitated through PRS, teacher identity and teaching approach including perceived enthusiasm, commitment and the effective management of students in class; authentic and constructively aligned curriculum and assessment; an appropriate mix of teacher guided and student centred learning, and clear alignment between the learning experience and outcomes and student's current and future professional practice.

As the literature identifies, these elements have been the subject of considerable research, including in relation to whether and how they may foster deep versus surface learning. Biggs (1987) identifies three dimensions that relate to student's motivation for learning. In doing so he differentiates between a deep, surface and achieving approach. When deep learning occurs students' motivations to learn are intrinsic. By contrast, when surface learning occurs students are motivated by fear, and a desire to get through with minimal effort. It is the achieving dimension which relates to the strategic decision making that occurs for the learner.

The degree to which there exists a direct correlation here between the pedagogic approach adopted in this course and deeper learning requires further targeted research. However, by drawing upon the literature and applying Tangey's lens of student centred learning this chapter argues that learner engagement, and we suggest deeper approaches to learning, were evidenced through student feedback which highlights the benefits of learner orientated teaching. It is through a learning orientated pedagogy that we may create the real potential of rich quality learning outcomes. This in turn supports effective preparation for professional practice (Kember & Leung, 2005b). As this study affirms, it is through fostering learning orientated active and student-centred learning that our students can become inspired to learn and achieve quality learning outcomes. Both elements are notably essential to fostering effective preparation for professional practice and to enabling students to conceptualise learning as lifelong (Kember, 2009).

The authors have analysed the student's perceptions of their learning journey within the large class context affirming the critical role of using active student centred learning strategies to facilitate learner engagement Our findings further highlight the importance of actively engaging learners to ensure that students develop the requisite knowledge and skills; and in doing so, value learning and the learning process.

We have argued here for the conceptualisation of learning as a meaning making process creating the potential for deeper learning to arise through pedagogies dedicated to ensuring learner engagement. Further research is needed to explore whether and how learner engagement in turn drives intrinsic motivation.

The learner centred pedagogy has a deep resonance for both the particular discipline of Project Management and for good practice more broadly in higher education. This importantly includes within the large class context of tertiary education.

REFERENCES

Baeten, M., Dochy, F., Struyven, K., Parmentier, E., & Vanderbruggen, A. (2016). Student-centred learning environments: An investigation into student teachers' instructional preferences and approaches to learning. *Learning Environments Research*, *19*(1), 43–62. doi:10.100710984-015-9190-5

Baeten, M., Kyndt, E., Struyven, K., & Dochy, F. (2010). Using student-centred learning environments to stimulate deep approaches to learning: Factors encouraging or discouraging their effectiveness. *Educational Research Review*, *5*(3), 243–260. doi:10.1016/j.edurev.2010.06.001

Bates, S. P., Howie, K. & Murphy, A. S. J. (2006). The use of electronic voting systems in large group lectures: challenges and opportunities. *New Directions*, 1-8.

Bentley, Y., Richardson, D., Duan, Y., Philpott, E., Ong, V., & Owen, D. (2013). Research-informed curriculum design for a master's-level program in project management. *Journal of Management Education*, *37*(5), 651–682. doi:10.1177/1052562912458642

Biggs, J. B. (1987). *Student Approaches to Learning and Studying. Research Monograph*. ERIC.

Biggs, J. B. (1999). *Teaching for Quality Learning at University*. SHRE and Open University Press.

Cope, C., & Staehr, L. (2005). Improving students' learning approaches through intervention in an information systems learning environment. *Studies in Higher Education, 30*(2), 181–197. doi:10.1080/03075070500043275

Entwistle, N. (1988). *Styles of Learning and Teaching*. David Fulton.

Gysbers, V., Johnston, J., Hancock, D., & Denyer, G. (2011). Why do students still bother coming to lectures, when everything is available online? 19(2), 20-36. *International Journal of Innovation in Science and Mathematics Education, 19*, 20–36.

Kember, D. (2009). Promoting student-centred forms of learning across an entire university. *Higher Education, 58*(1), 1–13. doi:10.100710734-008-9177-6

Kember, D., & Leung, D. Y. (2005a). The influence of active learning experiences on the development of graduate capabilities. *Studies in Higher Education, 30*(2), 155–170. doi:10.1080/03075070500043127

Kember, D., & Leung, D. Y. (2005b). The influence of the teaching and learning environment on the development of generic capabilities needed for a knowledge-based society. *Learning Environments Research, 8*(3), 245–266. doi:10.100710984-005-1566-5

Kember, D., Leung, D. Y., & Mcnaught, C. (2008). A workshop activity to demonstrate that approaches to learning are influenced by the teaching and learning environment. *Active Learning in Higher Education, 9*(1), 43–56. doi:10.1177/1469787407086745

Kember, D. (2004). Interpreting student workload and the factors which shape students' perceptions of their workload. *Studies in Higher Education, 29*, 165-184.

Lea, S. J., Stephenson, D., & Troy, J. (2003). Higher education students' attitudes to student-centred learning: Beyond 'educational bulimia'? *Studies in Higher Education, 28*(3), 321–334. doi:10.1080/03075070309293

Liu, D. Y., & Taylor, C. E. (2013). Engaging students in large lectures of introductory biology and molecular biology service courses using student response systems. *Proceedings of The Australian Conference on Science and Mathematics Education*.

Masikunas, G., Panayiotidis, A., & Burke, L. (2007). The Use of Electronic Voting Systems in Lectures within Business and Marketing: A Case Study of Their Impact on Student Learning. *ALT-J: Research in Learning Technology, 15*, 3-20.

Meltzer, D. E., & Manivannan, K. (2002). Transforming the lecture-hall environment: The fully interactive physics lecture. *American Journal of Physics, 70*(6), 639–654. doi:10.1119/1.1463739

Moss, K., & Crowley, M. (2011). Effective learning in science: The use of personal response systems with a wide range of audiences. *Computers & Education*, *56*(1), 36–43. doi:10.1016/j.compedu.2010.03.021

Nicol, D. J., & Macfarlane-Dick, D. (2006). Formative assessment and self-regulated learning: A model and seven principles of good feedback practice. *Studies in Higher Education*, *31*(2), 199–218. doi:10.1080/03075070600572090

O'Neill, G., & McMahon, T. (2005). *Student-centred learning: What does it mean for students and lecturers*. Academic Press.

Pant, I., & Baroudi, B. (2008). Project management education: The human skills imperative. *International Journal of Project Management*, *26*(2), 124–128. doi:10.1016/j.ijproman.2007.05.010

Ramsden, P. (1992). *Learning to Teach in Higher Education*. Routledge.

Surentheran, T. (2014). Use of an Electronic Voting System to facilitate interactive engagement and enhance active learning of undergraduate students in Biomedical Science. *Investigations in University Teaching and Learning, 9*, 111-121.

Tangney, S. (2014). Student-centred learning: A humanist perspective. *Teaching in Higher Education*, *19*(3), 266–275. doi:10.1080/13562517.2013.860099

Voelkel, S., & Bennett, D. (2014). New uses for a familiar technology: Introducing mobile phone polling in large classes. *Innovations in Education and Teaching International*, *51*(1), 46–58. doi:10.1080/14703297.2013.770267

Wilson, K., & Fowler, J. (2005). Assessing the impact of learning environments on students' approaches to learning: Comparing conventional and action learning designs. *Assessment & Evaluation in Higher Education*, *30*(1), 87–101. doi:10.1080/0260293042003251770

Wolter, B. H., Lundeberg, M. A., Kang, H., & Herreid, C. F. (2011). Students' perceptions of using personal response systems ("clickers") with cases in science. *Journal of College Science Teaching*, *40*, 14–19.

Compilation of References

Abdirad, H., & Dossick, C. S. (2016). BIM curriculum design in architecture, engineering, and construction education: A systematic review. *ITcon*, *21*, 250–271.

Abdirad, H., & Dossick, C. S. (2016). BIM curriculum design in architecture, engineering, and construction education: A systematic review. *Journal of Information Technology in Construction*, *21*(17), 250–271.

Abokhodair, E. S. (2015). Action learning as an approach for developing leaders, solving problems, and building successful teams: A case study of princess nora bint abdul-rahman university. In Community Engagement in Higher Education: Policy Reforms and Practice. Sense Publishers. doi:10.1007/978-94-6300-007-9_18

Ackoff, R. L. (1974). *Redesigning the Future: Systems Approach to Societal Problems*. Wiley.

Adwan, J. (2016). Dynamic online peer evaluations to improve group assignments in nursing e-learning environment. *Nurse Education Today*, *41*, 67–72. doi:10.1016/j.nedt.2016.03.028 PMID:27138485

Adya, M., Temple, B. K., & Hepburn, D. M. (2015). Distant yet Near: Promoting Interdisciplinary Learning in Significantly Diverse Teams through Socially Responsible Projects. *Decision Sciences Journal of Innovative Education*, *13*(2), 121–149. doi:10.1111/dsji.12058

Ahmed, A. K. (2013). Teacher centered versus learner centered teaching style. *The Journal of Global Business Management.*, *9*(1), 22–34.

Albeanu, G. (2015). Using Moodle for Teaching Science and Information Technology, Towards Higher Education Reengineering. *11th International Scientific Conference, eLearning and Software for Education*, 606-611.

Ali, M. F., Tahir, L. M., & Said, M. N. H. M. (2017). Facilitating learners' online discussions in completing group assignments. In *2016 IEEE Conference on e-Learning, e-Management and e-Services, IC3e 2016*. Institute of Electrical and Electronics Engineers Inc.

Ali, M. F., Tahir, L. M., Said, M. N. H. M., Junaidi, J., Atan, N. A., & Hahsan, A. (2018). Promoting staged self-directed learning (SSDL) among Malaysian tertiary learners through online discussion in completing group assignment. In *2017 IEEE International Conference on Teaching, Assessment and Learning for Engineering, TALE 2017*. Institute of Electrical and Electronics Engineers Inc.

Alshuwaikhat Habib, M. A. Y. (2016). Sustainability Assessment of Higher Education Instituions in Saudi Arabia. *MDPI, 8*.

Amory, A. (2012). Tool-Mediated Authentic Learning in an Educational Technology Course: A Design-Based Innovation. *Interactive Learning Environments, 22*(4), 497–513. doi:10.1080/10494820.2012.682584

Angelva, J., Tepsa, T., & Mielikäinen, M. (2017). Team teaching experiences in engineering education a project-based learning approach. In *45th Annual Conference of the European Society for Engineering Education, SEFI 2017*. European Society for Engineering Education (SEFI).

Arashpour, M., Bai, Y., Kamat, V., Hosseini, R., & Martek, I. (2018). Project production flows in off-site prefabrication: BIM-enabled railway infrastructure. In *35th International Symposium on Automation and Robotics in Construction and International AEC/FM Hackathon: The Future of Building Things, ISARC 2018*. International Association for Automation and Robotics in Construction I.A.A.R.C).

Arashpour, M., Miletic, M., Williams, N., & Fang, Y. (2018). Design for manufacture and assembly in off-site construction: Advanced production of modular façade systems. In *35th International Symposium on Automation and Robotics in Construction and International AEC/FM Hackathon: The Future of Building Things, ISARC 2018*. International Association for Automation and Robotics in Construction I.A.A.R.C).

Arashpour, M., & Aranda-Mena, G. (2017). Curriculum renewal in architecture, engineering, and construction education: Visualizing building information modeling via augmented reality. In *9th International Structural Engineering and Construction Conference: Resilient Structures and Sustainable Construction, ISEC 2017*. ISEC Press.

Arashpour, M., Bai, Y., Aranda-mena, G., Bab-Hadiashar, A., Hosseini, R., & Kalutara, P. (2017). Optimizing decisions in advanced manufacturing of prefabricated products: Theorizing supply chain configurations in off-site construction. *Automation in Construction, 84*, 146–153. doi:10.1016/j.autcon.2017.08.032

Arashpour, M., Kamat, V., Bai, Y., Wakefield, R., & Abbasi, B. (2018). Optimization modeling of multi-skilled resources in prefabrication: Theorizing cost analysis of process integration in off-site construction. *Automation in Construction, 95*, 1–9. doi:10.1016/j.autcon.2018.07.027

Arashpour, M., Sagoo, A., Wingrove, D., Maqsood, T., & Wakefield, R. (2015). Single capstone or multiple cornerstones? Distributed model of capstone subjects in construction education. In *Proceedings of 8th International Structural Engineering and Construction Conference, ISEC 2015*. ISEC Press.

Arashpour, M., Too, E., & Le, T. (2017). Improving productivity, workflow management, and resource utilization in precast construction. In *9th International Structural Engineering and Construction Conference: Resilient Structures and Sustainable Construction, ISEC 2017*. ISEC Press.

Arashpour, M., Wakefield, R., Abbasi, B., Arashpour, M., & Hosseini, R. (2018). Optimal process integration architectures in off-site construction: Theorizing the use of multi-skilled resources. *Architectural Engineering and Design Management*, *14*(1-2), 46–59. doi:10.1080/17452007.2017.1302406

Argyris, C. (1985). *Action Science, Concepts, methods, and skills for research and intervention.* San Francisco: Jossey-Bass.

Asok, D., Abirami, A. M., Angeline Cv, N., & Lavanya, R. (2017). Active learning environment for achieving higher-order thinking skills in engineering education. In *4th IEEE International Conference on MOOCs, Innovation and Technology in Education, MITE 2016*. Institute of Electrical and Electronics Engineers Inc.

Asper, N. (2010). *Instructions for poster presentations*. Retrieved 1 June, 2011, from http://www.tcnj.edu/~asper/posterrules.html

Awatramani, M., & Rover, D. (2015). Team-based learning course design and assessment in computer engineering. In *2015 IEEE Frontiers in Education Conference, FIE 2015*. Institute of Electrical and Electronics Engineers Inc.

Bada, S.O. (2015). Constructivism Learning Theory: A Paradigm for Teaching and Learning. *IOSR Journal of Research & Method in Education, 5*(6), 66-70.

Baeten, M., Dochy, F., Struyven, K., Parmentier, E., & Vanderbruggen, A. (2016). Student-centred learning environments: An investigation into student teachers' instructional preferences and approaches to learning. *Learning Environments Research*, *19*(1), 43–62. doi:10.100710984-015-9190-5

Baeten, M., Kyndt, E., Struyven, K., & Dochy, F. (2010). Using student-centred learning environments to stimulate deep approaches to learning: Factors encouraging or discouraging their effectiveness. *Educational Research Review*, *5*(3), 243–260. doi:10.1016/j.edurev.2010.06.001

Baker, N., & Standeven, M. (1996). Thermal comfort for free-running buildings. *Energy and Building*, *23*(3), 175–182. doi:10.1016/0378-7788(95)00942-6

Bargard, A., Corsaro, M., & Ewaney, M. J. (2014). The Poster Presentation as Experiential Education: Bridging the Gap between Research and Practice. *Clinical. The School Review*, *7*(2), 105–108. doi:10.1891/1939-2095.7.2.105

Barlish, K., & Sullivan, K. (2012). How to measure the benefits of BIM—A case study approach. *Automation in Construction*, *24*, 149–159. doi:10.1016/j.autcon.2012.02.008

Bates, S. P., Howie, K. & Murphy, A. S. J. (2006). The use of electronic voting systems in large group lectures: challenges and opportunities. *New Directions*, 1-8.

Becerik-Gerber, B., Gerber, D. J., & Ku, K. (2011). The pace of technological innovation in architecture, engineering, and construction education: Integrating recent trends into the curricula. *ITcon, 16*, 411–432.

Becerik-Gerber, B., Gerber, D. J., & Ku, K. (2011). The pace of technological innovation in architecture, engineering, and construction education: Integrating recent trends into the curricula. *Journal of Information Technology in Construction, 16*, 411–432.

Bentley, Y., Richardson, D., Duan, Y., Philpott, E., Ong, V., & Owen, D. (2013). Research-informed curriculum design for a master's-level program in project management. *Journal of Management Education, 37*(5), 651–682. doi:10.1177/1052562912458642

Bew, M., & Underwood, J. (n.d.). *Delivering BIM to the UK market.* doi:10.4018/978-1-60566-928-1.ch003

Beylefeld, A., & Le Roux, A. (2015). Guided group reflections of first-year pre-service teachers: Moving beyond the rhetoric of "go and reflect". *Perspectives in Education, 33*(3), 1–19.

Biggs, J. (1999). Enhancing teaching through constructive alignment. *Higher Education, 32*(3), 347–364. doi:10.1007/BF00138871

Biggs, J. (2007). *Teaching for Quality Learning in University.* Open University Press.

Biggs, J. B. (1987). *Student Approaches to Learning and Studying. Research Monograph.* ERIC.

Biggs, J. B. (2011). *Teaching for quality learning at university: What the student does.* McGraw-Hill Education.

BIM2050 Group. (2014). *Built Environment 2050 – A Report on Our Digital Future.* London: Construction Industry Council.

Birkenholz, R. (1999). *Effective adult learning.* Danville, IL: Interstate Publishers, Inc.

Black, P. (2003c). *Formative and summative assessment: Can they serve learning together?* Paper presented at AERA, SIG Classroom Assessment Meeting.

Black, P., Harrison, C., Lee, C., Marshall, B., & William, D. (2004). *Assessment for Learning. Putting it into Practice.* Maidenhead, UK: Open University Press.

Bloom, B. S. (1968). *Learning for mastery.* Los Angeles, CA: University of California Press.

Bornehag, C. G., Sundell, J., Bonini, S., Custovic, A., Malmberg, P., Skerfving, S., ... Verhoeff, A. (2004). Dampness in buildings as a risk factor for health effects, EUROEXPO: A multidisciplinary review of the literature (1998–2000) on dampness and mite exposure in buildings and health effects. *Indoor Air, 14*(4), 243–257. doi:10.1111/j.1600-0668.2004.00240.x PMID:15217478

Botterill, M., & Harpe, B. (2010). Working successfully in university interdisciplinary teams: Learning from embedded intergroup relations theory. International Perspectives on Higher Education Research, 5, 75-96.

Brandes, D., & Ginnis, P. (1996). *A Guide to Student Centred Learning*. Simon & Schuster.

Brewer, P. E., Mitchell, A., Sanders, R., Wallace, P., & Wood, D. D. (2015). Teaching and learning in cross-disciplinary virtual teams. *IEEE Transactions on Professional Communication*, *58*(2), 208–229. doi:10.1109/TPC.2015.2429973

Briggs, J., & Tang, C. (2007). *Teaching for Quality Learning at University*. New York: McGraw Hill.

Briscoe. (1996). *Preparing Scientific Information*. Springer-Verlag.

Brooks, N., & Moriarty, A. (2006). Development of a practice learning team in the clinical setting. Nursing Standard, 20(33), 41-44.

Brown, G., Bull, J., & Pendlebury, M. (1997). *Assessing student learning in higher education*. London: Routledge.

Bruner, J. (1970). *The Process of Education*. Cambridge, MA: Harvard University Press.

Bryman, A., & Bell, E. (2011). *Business Research Methods* (3rd ed.). Oxford University Press.

Bui, H. T. M., Baruch, Y., Chau, V. S., & He, H. W. (2016). Team learning: The missing construct from a cross-cultural examination of higher education. *Asia Pacific Journal of Management*, *33*(1), 29–51. doi:10.100710490-015-9426-z

Burgess Yakemovic, K. C., & Conklin, E. J. (1990). Report on a development project use of an issue-based information system. In *Proceedings of the 1990 ACM conference on Computer-supported cooperative work - CSCW '90* (pp. 105–118). New York: ACM Press. 10.1145/99332.99347

Burgess-Allen, J., & Owen-Smith, V. (2010). Using mind mapping techniques for rapid qualitative data analysis in public participation processes. *Health Expectations : An International Journal of Public Participation in Health Care and Health Policy*, *13*(4), 406–415. doi:10.1111/j.1369-7625.2010.00594.x PMID:20550595

Cabinet Office. (2011). *Government Construction Strategy*. Available from: https://www.gov.uk/government/uploads/system/uploads/attachment_data/file/61152/Government-Construction-Strategy_0.pdf

Cannon, R., & Newble, D. (2000). *A Guide to Improving Teaching Methods: a handbook for teachers in universities and colleges*. Kogan Page.

Caple, H., & Bogle, M. (2013). Making group assessment transparent: What wikis can contribute to collaborative projects. *Assessment & Evaluation in Higher Education*, *38*(2), 198–210. doi:10.1080/02602938.2011.618879

CESE. (n.d.). *Developing a logic model*. Retrieved from https://education.nsw.gov.au/teaching-and-learning/professional-learning/evaluation-resource-hub/media/documents/Develop-a-logic-model-presentation_v2.pdf

Chapman, K. J., Meuter, M., Toy, D., & Wright, L. (2006). Can't we pick our own groups? The influence of group selection method on group dynamics and outcomes. *Journal of Management Education, 30*(4), 557–569. doi:10.1177/1052562905284872

Chegu Badrinath, A., Chang, Y. T., & Hsieh, S. H. (2016). A review of tertiary BIM education for advanced engineering communication with visualisation. *Visualisation in Engineering*, *4*(1), 9. doi:10.118640327-016-0038-6

Chen, C. C., Wu, J., & Hayashi, A. M. (2006). Importance of diversified leadership roles in improving team effectiveness in a virtual collaboration learning environment. *12th Americas Conference on Information Systems, AMCIS 2006.*

Cheng, R. (2006). Suggestions for an Integrative Education. In N. Strong, D. Friedman, & M. Broshar (Eds.), *Report on Integrated Practice*. Washington, DC: AIA.

Churchman, C. W. (1967). Wicked Problems. *Management Science*, 141–146. doi:10.1287/mnsc.14.4.B141

Clapp, A. (2017). An e-learning team's life on and offline: A collaborative self-ethnography in postgraduate education development. *Electronic Journal of e-Learning, 15*(1), 33-45.

Coccoli, M., Stanganelli, L., & Maresca, P. (2011). Computer Supported Collaborative Learning in software engineering: (ETC - Enforcing Team cooperation). *2011 IEEE Global Engineering Education Conference, EDUCON 2011.* 10.1109/EDUCON.2011.5773267

Coghlan, D. (2007). Insider action research: Opportunities and challenges. *Management Research News, 30*(5), 335–343. doi:10.1108/01409170710746337

Coghlan, D., & Brannick, T. (2014). *Doing Action Research in Your Own Organization*. Sage Publications London.

Conklin, J. (2005). *Dialogue Mapping: Building Shared Understanding*. Wiley.

Conte, E. (2013). Sustainability and built environment: The role of higher education in architecture and building engineering. *European Journal of Sustainable Development*, *5*(3), 1–10.

Cope, C., & Staehr, L. (2005). Improving students' learning approaches through intervention in an information systems learning environment. *Studies in Higher Education, 30*(2), 181–197. doi:10.1080/03075070500043275

Cotana, F., Rossi, F., Filipponi, M., Coccia, V., Pisello, A. L., Bonamente, E., ... Cavalaglio, G. (2014). Albedo control as an effective strategy to tackle Global Warming: A case study. *Applied Energy, 130*, 641–647. doi:10.1016/j.apenergy.2014.02.065

Cowan, J. (2005). *Student-centered learning*. Paper presented at DIT Learning and Teaching Showcase, Dublin Institute of Technology.

Crawford, R., & Jenkins, L. (2017). Blended learning and team teaching: Adapting pedagogy in response to the changing digital tertiary environment. *Australasian Journal of Educational Technology*, *33*(2), 51–72.

Cross, K. P. (1996). Improving teaching and learning through classroom assessment and classroom research. In Improving student learning using research to improve student learning. Academic Press.

Crowe, M., & Sheppard, L. (2011). Mind mapping research methods. *Quality & Quantity*, *46*(5), 1493–1504. doi:10.100711135-011-9463-8

Dacre Pool, L., & Sewell, P. (2007). The key to employability: Developing a practical model of graduate employability. *Education + Training*, *49*(4), 277–289. doi:10.1108/00400910710754435

de la Harpe, B., Kulski, M., & Radloff, A. (1999). How best to document the quality of our teaching and our students' learning? In K. Martin, N. Stanley, & N. Davison (Eds.), Teaching in the Disciplines/ Learning in Context (pp. 108–113). Academic Press.

Debuse, J. C. W., & Lawley, M. (2016). Benefits and drawbacks of computer-based assessment and feedback systems: Student and educator perspectives. *British Journal of Educational Technology*, *47*(2), 294–301. doi:10.1111/bjet.12232

Denson, N., & Zhang, S. (2010). The impact of student experiences with diversity on developing graduate attributes. *Studies in Higher Education*, *35*(5), 529–543. doi:10.1080/03075070903222658

Deonandran, R., Gomes, J., Lavigne, E., & Blanchard, R. (2013). A pilot study: research poster presentations as an education tool for undergraduate epidemiology students. *Advances in Medical Education and Practice, 4*, 183-188.

Deutsch, R. (2011). *BIM and Integrated Design: Strategies for Architectural Practice*. John Wiley & Sons, Inc.

Dicheva, D., Dicheva, C., Agre, G., & Angelova, G. (2015). Gamification in Education: A Systematic Mapping Study. *Journal of Educational Technology & Society*, *18*(3), 75–88.

Donaldson, M. (1987). The origins of inference. In J. Bruner & H. Haste (Eds.), *Making Sense. The Child's Construction of the World*. London: Routledge.

Ecclestone, K. (1999). Empowering or Ensnaring? The Implications of Outcome-based Assessment in Higher Education. *Higher Education Quarterly*, *53*(1), 29–48. doi:10.1111/1468-2273.00111

Elahi, B. (2017). Implementation of a learning assessment: Case study of applied statistics course in industrial engineering. In *124th ASEE Annual Conference and Exposition*. American Society for Engineering Education. 10.18260/1-2--27927

Engelbrecht, F., (2017). *Detailed projections of future climate change over South Africa*. CSIR Technical Report.

Entwistle, N. (1988). *Styles of Learning and Teaching*. David Fulton.

Ewell, P. (2007). *Applying learning outcomes concepts to higher education: An overview*. Prepared for the Hong Kong University Grants Committee.

Farrington, J. (1991). Student-centred learning: Rhetoric and reality*? Journal of Further and Higher Education, 15*(3), 16–21. doi:10.1080/0309877910150302

Fazey, D. (1996). Individual differences in the autonomy related psychological characteristics of undergraduates. In G. Gibbs (Ed.), *Improving student learning using research to improve student learning* (pp. 187–199). Oxford, UK: Oxford Centre for Staff Development.

Feagin, J., Orum, A., & Sjoberg, G. (Eds.). (1991). *A case for case study*. Chapel Hill, NC: University of North Carolina Press.

Feldmann, B. (2006). Group types in E-learning environments - Study team, working team and learning team. In *7th International Conference on Information Technology Based Higher Education and Training*. ITHET.

Ficano, C. K. C. (2018). Identifying differential benefits from a flipped-group pedagogy in introductory microeconomics. *International Review of Economics Education.*

Finkel, D., & Monk, G. (1983). Teachers and Learning Groups: Dissolution of the Atlas Complex. In C. Bouton & R. Y. Garth (Eds.), *Learning in Groups. New Directions for Teaching and Learning*. Jossey-Bass. doi:10.1002/tl.37219831411

Fisk, W. J., Lei-Gomez, Q., & Mendell, M. J. (2007). Meta-analyses of the associations of respiratory health effects with dampness and mold in homes. *Indoor Air, 17*(4), 284–296. doi:10.1111/j.1600-0668.2007.00475.x PMID:17661925

Forgues, D., & Iordanova, I. (2010). An IDP-BIM Framework for Reshaping Professional Design Practices. In *Construction Research Congress 2010* (pp. 172-182). American Society of Civil Engineers. 10.1061/41109(373)18

Fosnot, C. T. (1996). *Constructivism: Theory, Perspectives and Practice*. New York: College Teacher Press.

Francis, K., Henderson, M., Martin, E., Saul, K., & Joshi, S. (2018). Collaborative teaching and interdisciplinary learning in graduate environmental studies. *Journal of Environmental Studies and Sciences, 8*(3), 343–350. doi:10.100713412-017-0467-0

Francois, J.-C. B. A., Blackwood, D. J., & Jowitt, P. W. (2002). Decision mapping : Understanding decision making processes. *Civil Engineering and Environmental Systems, 19*(3), 187–207. doi:10.1080/10286600214151

Freemana, S., Eddya, S. L., McDonougha, M., Smith, M. K., Okoroafora, M., Jordta, J., & Wenderotha, M. (2014). Active learning increases student performance in science, engineering, and mathematics. PNAS, Psychological and Cognitive Sciences, 111(23).

Fry, H., Ketteridge, S., & Marshall, S. (1999) Understanding student learning. In A Handbook for Teaching and Learning in higher Education. Enhancing Academic Practice. London: Kogan Page.

Fuji, T., Tanigawa, T., Inui, M., & Saegusa, T. (1996). CAMELOT: Collaborative and multimedia environment for learners on teams. *Education and Information Technologies*, *1*(3-4), 203–226. doi:10.1007/BF02350659

Funnell, S. C., & Rogers, P. J. (2011). *Purposeful Program Theory:Effective Use of Theories of Change and Logic Models*. Hoboken, NJ: Wiley.

Gabelnick, F., MacGregor, J., Matthews, R., & Smith, B. (1990). *Learning Communities: Creating Connections Among Students, Faculty and Disciplines.* San Francisco: Jossey Bass.

Gerber, D., Khashe, S., & Smith, I. (2013). Surveying the Evolution of Computing in Architecture, Engineering, and Construction Education. *Journal of Computing in Civil Engineering*, (Oct): 7.

Gibberd, J. (2009a). Water Conservation. In Green Building Handbook (vol. 1). Alive 2 Green Publishers.

Gibberd, J. (2009b). Building Envelope. In Green Building Handbook (vol. 1). Alive 2 Green Publishers.

Gibberd, J., (2015). Rainwater harvesting playing a valuable role in increasing the resilience and sustainability of water supply. In *Sustainable Water Resource Handbook* (vol. 6). Alive 2 Green Publishers.

Gibberd, J. (2017a). Water Resilience in Urban Areas. *11th Built Environment Conference*, Durban, South Africa.

Gibberd, J. (2017b). Social Resilience in Urban Areas. *11th Built Environment Conference*, Durban, South Africa.

Gibbs, G. (1992). *Improving the quality of student learning*. Bristol, UK: TES.

Gledson, B., Hilton, D., & Rogage, K. (2016). Benchmarking BIM Levels of Training and Education amongst Construction Management Practitioners. *Proceedings of the 32nd Annual ARCOM Conference*.

González, C., & Blanco, F. (2008). Integrating an Educational 3D Game in Moodle. *Simulation & Gaming*, *39*(3), 399–413. doi:10.1177/1046878108319585

Goodman, H., Knight, C., & Khudododov, K. (2014). Graduate Social Work Students' Experiences with Group Work in the Field and the Classroom. *Journal of Teaching in Social Work*, *34*(1), 60–78. doi:10.1080/08841233.2013.866615

Gordon, C., Azambuja, M., & Werner, A. M. (2009). BIM across the construction curriculum. *Proc. of the 2009 ASC Region III Conference*.

Gore, J., & Zeichner, K. (1991). Action research and reflective teaching in preservice teacher education – a case study from the United States. *Teaching and Teacher Education*, *7*(2), 119–136. doi:10.1016/0742-051X(91)90022-H

Gosling, P. J. (1999). *Scientist's guide to Poster Presentations.* New York: Springer LLC. doi:10.1007/978-1-4615-4761-7

Goth, U. S., Bergsli, O., & Johanesen, E. M. (2017). Integration of ethnic minorities during group-work for vocational teachers-in-training in health studies. *International Journal of Medical Education, 8*, 31–36. doi:10.5116/ijme.5862.505a PMID:28132033

Government, N. S. W. (2016). *NSW Government Program Evaluation Guidelines.* Sydney: NSW Department of Premier and Cabinet.

Grabinger, R. S. (1996). Rich environments for active learning. In D. H. Jonassen (Ed.), *Handbook for research for educational communications and technology.* New York: MacMillan.

Grunert O'Brien, J., Millis, B., & Cohen, M. (2009). *The Course Syllabus: A Learning-Centered Approach.* John Wiley and Sons.

Guan, L. (2009). Preparation of future weather data to study the impact of climate change on buildings. *Building and Environment, 44*(4), 793–800. doi:10.1016/j.buildenv.2008.05.021

Guilbaud, P., & Whitney, M. (2018). Technology-enhanced learning and the multi-cultural classroom: Exploring impacts of open educational resources' use on collaboration and teamwork. In *6th International Conference on Information and Communication Technology and Accessbility, ICTA 2017.* Institute of Electrical and Electronics Engineers Inc.

Gysbers, V., Johnston, J., Hancock, D., & Denyer, G. (2011). Why do students still bother coming to lectures, when everything is available online? 19(2), 20-36. *International Journal of Innovation in Science and Mathematics Education, 19*, 20–36.

Hallegatte, S. (2009). Strategies to adapt to an uncertain climate change. *Global Environmental Change, 19*(2), 240–247. doi:10.1016/j.gloenvcha.2008.12.003

Hamin, E. M., & Gurran, N. (2009). Urban form and climate change: Balancing adaptation and mitigation in the US and Australia. *Habitat International, 33*(3), 238–245. doi:10.1016/j.habitatint.2008.10.005

Hampton, D. C., & El-Mallakh, P. (2017). Opinions of online nursing students related to working in groups. *The Journal of Nursing Education, 56*(10), 611–616. doi:10.3928/01484834-20170918-06 PMID:28972631

Han, S. J., Liau-Hing, C., & Beyerlein, M. (2016). Facilitating multicultural student team engagement in higher education: A model for digital learning environments. In Handbook of Research on Transformative Digital Content and Learning Technologies. IGI Global.

Handbook of Research on Building Information Modelling and Construction Informatics. (2010). *Concepts and Technologies.* Hershey, PA: IGI Global.

Hannele, K., Reijo, M., Tarja, M., Sami, P., Jenni, K., & Teija, R. (2012). Expanding uses of building information modeling in life-cycle construction projects. *Work (Reading, Mass.), 41*, 114–119. PMID:22316709

Harden, R. M., & Crosby, J. (2000). AMEE Guide No 20: The good teacher is more than a lecturer-the twelve roles of the teacher. *Medical Teacher*, *22*(4), 334–347. doi:10.1080/014215900409429

Harding, L. M. (2018). Students of a Feather "Flocked" Together: A Group Assignment Method for Reducing Free-Riding and Improving Group and Individual Learning Outcomes. *Journal of Marketing Education*, *40*(2), 117–127. doi:10.1177/0273475317708588

Harlen, W., & James, M. (1997). Assessment and learning: Differences and relationships between formative and summative assessment. *Assessment in Education: Principles, Policy & Practice*, *4*(3), 365–379. doi:10.1080/0969594970040304

Hartley, J. (1998). *Learning and Studying. A research perspective*. London: Routledge. doi:10.4324/9780203270189

Harvey, L. (2003). *Transitions from higher education to work: a briefing paper prepared by Lee Harvey (Centre for Research and Evaluation, Sheffield Hallam University), with advice from ESECT and LTSN Generic Centre colleagues*. Sheffield Hallum University.

Hatvani-Kovacs, G., Belusko, M., Skinner, N., Pockett, J., & Boland, J. (2016). Heat stress risk and resilience in the urban environment. *Sustainable Cities and Society*, *26*, 278–288. doi:10.1016/j.scs.2016.06.019

Henderson, L., & Jordan, N.L. (2009, Fall). A Modest Proposal for a Transdisciplinary Curriculum for the Design, Construction, Management and Maintenance of Architecture. *Journal of Building Information Modeling*.

Herr, C. M. (2013). Qualitative structural design education in large cohorts of undergraduate architecture students. *Global Journal of Engineering Education*, *15*(2), 96–102.

Herrington, J., & Parker, J. (2013). Emerging Technologies as Cognitive Tools for Authentic Learning. *British Journal of Educational Technology*, *44*(4), 607–615. doi:10.1111/bjet.12048

Hess, G., Tosney, K., & Liegel, L. (2011). *Creating Effective Poster Presentations: An Effective Poster*. Retrieved 1 June, 2011, from http://www.ncsu.edu/project/posters

Hessler, K. L., & Henderson, A. M. (2013). Interactive learning research: Application of cognitive load theory to nursing education. *International Journal of Nursing Education Scholarship*, *10*(1), 1–9. doi:10.1515/ijnes-2012-0029 PMID:23813334

Hewitson, B. C., & Crane, R. G. (2006). Consensus between GCM climate change projections with empirical downscaling: Precipitation downscaling over South Africa. *International Journal of Climatology*, *26*(10), 1315–1337. doi:10.1002/joc.1314

Hogle, L. (2013). *Industry + Academia: The perfect partnership*. Paper presented at the BIM Academic Education Symposium. Setting the Course for a BIM Educational Strategy, Washington, DC.

Horn, R. E. (2001). Knowledge Mapping for Complex Social Messes. *Knowledge Creation Diffusion Utilization*, 1–12. Retrieved from http://www.stanford.edu/~rhorn/SpchPackard.html

Howe, T. H., Sheu, C. F., & Hinojosa, J. (2018). Teaching theory in occupational therapy using a cooperative learning: A mixed-methods study. *Journal of Allied Health*, *47*(1), 66–71. PMID:29504022

Huff, P. L. (2014). The Goal Project: A Group Assignment to Encourage Creative Thinking, Leadership Abilities and Communication Skills. *Accounting Education*, *23*(6), 582–594. doi:10.1080/09639284.2014.974198

Hu, H. (2015). Building virtual teams: Experiential learning using emerging technologies. *E-Learning and Digital Media*, *12*(1), 17–33. doi:10.1177/2042753014558373

Hui, K. P., Zarei, H., & Duffield, C. (2015). How do we instil project management experience into young engineers? The use of posters as a learning tool. *Proceedings of the Australian Association for Engineering Education 2015.*

Hui, K. P., & Zarei, H. (2016). Using inquiry-based visual tools to aid inclusiveness in vocational education. *The 25th National Vocational Education and Training Research (NCVER) Conference.*

Hussey, T., & Smith, P. (2002). The trouble with learning outcomes. *Active Learning in Higher Education*, *3*(3), 220–233. doi:10.1177/1469787402003003003

Hu, Z. Z., Tian, P. L., Li, S. W., & Zhang, J. P. (2017). BIM-based integrated delivery technologies for intelligent MEP management in the operation and maintenance phase. *Advances in Engineering Software*. doi:10.1016/j.advengsoft.2017.08.007

Hyldegård, J., Hertzum, M., & Hansen, P. (2015). Studying collaborative information seeking: Experiences with three methods. In Collaborative Information Seeking: Best Practices, New Domains and New Thoughts. Springer International Publishing.

IPCC. (2007). *Climate change 2007: Synthesis report, fourth assessment report.* Cambridge, UK: International Panel on Climate Change and Cambridge University Press.

Jackson, D. (2016). Re-conceptualising graduate employability: The importance of pre-professional identity. *Higher Education Research & Development*, *35*(5), 925–939. doi:10.1080/07294360.2016.1139551

Jackson, K. M., & Trochim, W. M. K. (2002). Concept Mapping as an Alternative Approach for the Analysis of Open-Ended Survey Responses. *Organizational Research Methods*, *5*(4), 307–336. doi:10.1177/109442802237114

Jaques, P. A., & Viccari, R. M. (2006). Considering students' emotions in computer-mediated learning environments. In Z. Ma (Ed.), *Web-based intelligent e-learning systems: Technologies and applications* (pp. 122–138). Hershey, PA: Information Science Publishing. doi:10.4018/978-1-59140-729-4.ch006

Johnson, D., & Johnson, R. (1984) Structuring Groups for Cooperative Learning. *Journal of Management Education*, *9*(4).

Joshua, E., & Mariajose, C. (2013). *A comparison of peer evaluation methods in capstone design. 120th ASEE Annual Conference and Exposition*, Atlanta, GA.

Juran, J. M. (Ed.). (1974). Quality Control Handbook (3rd ed.). McGraw-Hill Book Company.

Kalfa, S., & Taksa, L. (2017). Employability, managerialism, and performativity in higher education: A relational perspective. *Higher Education*, *74*(4), 687–699. doi:10.100710734-016-0072-2

Kampe, J., & Oppliger, D. E. (2011). Effectiveness of team-based STEM project learning to recruit minority high school students to STEM. In *ASEE Annual Conference and Exposition, Conference Proceedings*. American Society for Engineering Education.

Karimpour, M., Belusko, M., Xing, K., Boland, J., & Bruno, F. (2015). Impact of climate change on the design of energy efficient residential building envelopes. *Energy and Building*, *87*, 142–154. doi:10.1016/j.enbuild.2014.10.064

Kauchak, D. P., & Eggen, P. D. (1998). *Learning and Teaching: Research Based Methods*. Boston: Allyn and Bacon.

Keevers, L., Lefoe, G., Leask, B., Sultan, F. K. P. D., Ganesharatnam, S., Loh, V., & Lim, J. S. Y. (2014). 'I like the people I work with. Maybe I'll get to meet them in person one day': Teaching and learning practice development with transnational teaching teams. *Journal of Education for Teaching*, *40*(3), 232–250. doi:10.1080/02607476.2014.903024

Kember, D. (2004). Interpreting student workload and the factors which shape students' perceptions of their workload. *Studies in Higher Education, 29*, 165-184.

Kember, D. (1997). A re-conceptualisation of the research into university academics conceptions of teaching. *Learning and Instruction*, *7*(3), 255–275. doi:10.1016/S0959-4752(96)00028-X

Kember, D. (2009). Promoting student-centred forms of learning across an entire university. *Higher Education*, *58*(1), 1–13. doi:10.100710734-008-9177-6

Kember, D., & Leung, D. Y. (2005a). The influence of active learning experiences on the development of graduate capabilities. *Studies in Higher Education*, *30*(2), 155–170. doi:10.1080/03075070500043127

Kember, D., & Leung, D. Y. (2005b). The influence of the teaching and learning environment on the development of generic capabilities needed for a knowledge-based society. *Learning Environments Research*, *8*(3), 245–266. doi:10.100710984-005-1566-5

Kember, D., Leung, D. Y., & Mcnaught, C. (2008). A workshop activity to demonstrate that approaches to learning are influenced by the teaching and learning environment. *Active Learning in Higher Education*, *9*(1), 43–56. doi:10.1177/1469787407086745

Kemmis, S., & McTaggart, R. (2000). *The action research planner* (3rd ed.). Deakin University.

Khosa, D. K., & Volet, S. E. (2013). Promoting effective collaborative case-based learning at university: A metacognitive intervention. *Studies in Higher Education*, *38*(6), 870–889. doi:10.1080/03075079.2011.604409

Khuzwayo, M. E. (2018). Assessment of group work in initial teacher education and training. *South African Journal of Education*, *38*(2).

Kim, K. P. (2015). *Conceptual Building Information Modelling Framework for Whole-House Refurbishment Based on LCC and LCA* (Unpublished doctoral dissertation). Aston University, Birmingham, UK.

Kim, K. P., & Park, K. S. (2018). Delivering value for money with BIM-embedded housing refurbishment. Facilities, 1-18. doi:10.1108/F-05-2017-0048

Kim, K.P., Ma, T., Baryah, A.S., Zhang, C., & Hui, K.M. (2016). Investigation of readiness for 4D and 5D BIM adoption in the Australian construction industry. *Management Review: An International Journal, 11*(2), 43-64.

Kim, K.P., Ochoa, J.J., & Kwon, D. (2018). Feasibility study of BIM integration with big data for construction scheduling. *Journal of Advanced Research in Dynamical and Control Systems, S11*, 1260-1265.

Kingsford, C., & Salzberg, S. L. (2008). What are decision trees? *Nature Biotechnology*, *26*(9), 1011–1013. doi:10.1038/nbt0908-1011 PMID:18779814

Kocaturk, T., & Kiviniemi, A. (2013). Challenges of integrating BIM in architectural education. Paper presented at the eCAADe 2013: Computation and Performance, Delft, The Netherlands.

Kocaturk, T., & Kiviniemi, A. (2013). Challenges of Integrating BIM in Architectural Education. *Proceedings of the 31st International Conference on Education and research in Computer Aided Architectural Design in Europe*.

Kolb, D. (1994). *Experiential Learning: Experience as the Source of Learning and Development*. Englewood Cliffs, NJ: Prentice-Hall.

Konopacki, S., Gartland, L., Akbari, H., & Rainer, L. (1998). *Demonstration of energy savings of cool roofs. Executive summary (No. LBNL-40673-Exec. Summ.). Lawrence Berkeley National Lab.* Environmental Energy Technologies Div. doi:10.2172/296884

Krathwohl, D. R. (2002). A Revision of Bloom's Taxonomy: An Overview. Theory Into Practice, 41(4).

Krathwohl, D. R., Bloom, B. S., & Masia, B. B. (1964). *Taxonomy of educational objectives: The classification of educational goals. In Handbook II: The affective domain.* New York: David McKay.

Krecar, I. M., Kolega, M., & Kunac, S. F. (2014). The Effects of Drinking Water on Attention. *Procedia: Social and Behavioral Sciences*, *159*, 577–583. doi:10.1016/j.sbspro.2014.12.428

Kugbeadjor, W., Suresh, S., & Renukappa, S. (2015). BIM awareness and readiness of postgraduate built environment students in West Midlands Universities, UK. In *CIB International Conference Proceedings. Going North for Sustainability: Leveraging Knowledge and Innovation for Sustainable Construction and Development* (pp. 531-543). London: IBEA Publications.

Lalopa, J. M., Jacobs, J. W. Jr, & Hu, C. (1999). Evolving learning environments: Another team-based example in higher education. *Journal of Hospitality & Tourism Education*, *10*(4), 30–37. doi:10.1080/10963758.1999.10685207

Lambert, S. C., Carter, A. J., & Lightbody, M. (2014). Taking the guesswork out of assessing individual contributions to group work assignments. *Issues in Accounting Education*, *29*(1), 169–180. doi:10.2308/iace-50637

Lapisa, R., Bozonnet, E., Abadie, M. O., & Salagnac, P. (2013). Cool roof and ventilation efficiency as passive cooling strategies for commercial low-rise buildings–ground thermal inertia impact. *Advances in Building Energy Research*, *7*(2), 192–208. doi:10.1080/17512549.2013.865559

Lea, S. J., Stephenson, D., & Troy, J. (2003). Higher education students' attitudes to student-centred learning: Beyond the 'educational bulimia'? *Studies in Higher Education*, *28*(3), 321–334. doi:10.1080/03075070309293

Lee, H. J., Kim, H., & Byun, H. (2017). Are high achievers successful in collaborative learning? An explorative study of college students' learning approaches in team project-based learning. *Innovations in Education and Teaching International*, *54*(5), 418–427. doi:10.1080/14703297.2015.1105754

Lee, N., & Hollar, D. A. (2013). Probing BIM education in construction engineering and management programs using industry perceptions. In *49th ASC Annual Int. Conf. Proc.* California Polytechnic State Univ.

Leite, F., & Wang, L. (2014). Process-Oriented Approach of Teaching Building Information Modeling in Construction Management. *Journal of Professional Issues in Engineering Education and Practice*, *140*(4), 4014004. doi:10.1061/(ASCE)EI.1943-5541.0000203

Lewin, K. (1935). *A Dynamic Theory of Personality*. New York: McGraw Hill.

Lewin, K. (1948). *Resolving social conflicts: selected papers on group dynamics*. University of Michigan.

Lieu, D. K., & Sorby, S. (2009). *Visualization, Modelling and Graphics for Engineering Design* (1st ed.). Cengage Learning.

Liu, D. Y., & Taylor, C. E. (2013). Engaging students in large lectures of introductory biology and molecular biology service courses using student response systems. *Proceedings of The Australian Conference on Science and Mathematics Education.*

Lu, J., & Law, N. (2012). Understanding Collaborative Learning Behaviour from Moodle Log Data. *Interactive Learning Environments*, *20*(5), 451–466. doi:10.1080/10494820.2010.529817

Mac Gregor, J. (1991). *What Difference Do Learning Communities Make?* Olympia, WA: Washington Center for Undergraduate Education, The Evergreen State College.

Macdonald, J. A. (2012). *A framework for collaborative BIM education across the AEC disciplines.* Paper presented at the 37th Annual Conference of Australasian University Building Educators Association (AUBEA).

MacGregor, J. (1990). Collaborative Learning: Shared Inquiry as a Process of Reform. In M. Svinicki (Ed.), *The Changing Face of College Teaching. New Directions for Teaching and Learning, No. 42*. San Francisco: Jossey- Bass. doi:10.1002/tl.37219904204

Maghool, Hossein (Iradj) Moeini, & Arefazar. (2018). An educational application based on virtual reality technology for learning architectural details: challenges and benefits. *International Journal of Architectural Research, 12*(3), 246-272.

Magiera, J. (2013). Integrated teaching for integrated engineering practice-BIM on campus. Paper presented at 210th International Conference on Computer Methods in Mechanics, Poznan, Poland.

Magogwe, J. M., Ntereke, B., & Phetlhe, K. R. (2015). Facebook and classroom group work: A trial study involving University of Botswana Advanced Oral Presentation students. *British Journal of Educational Technology, 46*(6), 1312–1323. doi:10.1111/bjet.12204

Maina, E. M., Oboko, R. O., & Waiganjo, P. W. (2017). *Extending moodle grouping functionality using artificial intelligent techniques. In IEEE AFRICON 2017.* Institute of Electrical and Electronics Engineers Inc.

Major, C. M., & Palmer, B. (2006). Reshaping teaching and learning: The transformation of faculty pedagogical content knowledge. *Higher Education, 51*(4), 619–647. doi:10.100710734-004-1391-2

Mangum, S. W., Johnson, J. L., Gunaldo, T. P., Casanova, E., & DeKerlegand, A. (2017). Developing student interprofessional competence through a support group assignment. *Journal of Interprofessional Education and Practice, 9*, 17–20. doi:10.1016/j.xjep.2017.07.006

Maqbool, J. (2016). Group work and the impact, if any, of the use of google applications for education. *International Conference on Cognition and Exploratory Learning in Digital Age.*

Mark, E., Martens, B., & Oxman, R. (2001). The Ideal Computer Curriculum. Architectural Information Management: eCAADe Helsinki, 168-175.

Masikunas, G., Panayiotidis, A., & Burke, L. (2007). The Use of Electronic Voting Systems in Lectures within Business and Marketing: A Case Study of Their Impact on Student Learning. *ALT-J: Research in Learning Technology, 15*, 3-20.

Masters, J. (1995). The history of action research. *Action Research.*

Mastropieri, M., & Scruggs, T. (2004). The Inclusive Classroom: Strategies for The Inclusive Classroom: Strategies for Effective Instruction Effective Instruction (2nd ed.). Pearson.

Matveev, A. V., & Milter, R. G. (2010). An implementation of active learning: Assessing the effectiveness of the team infomercial assignment. *Innovations in Education and Teaching International*, *47*(2), 201–213. doi:10.1080/14703291003718935

McCarthy, B. (1990). Using the 4MAT System to Bring Learning Styles to Schools. *Educational Leadership*, *48*(2), 31-37.

McGraw Hill. (2009). *Smart Market Report: The Business Value of BIM: Getting Building Information Modeling to the Bottom Line*. Bedford, MA: McGraw Hill Construction.

McGraw Hill. (2012). *Smart Market Report: The Business Value of BIM in North America: Multi-Year Trend Analysis and User Ratings (2007-2012)*. Bedford, MA: McGraw Hill Construction.

Meltzer, D. E., & Manivannan, K. (2002). Transforming the lecture-hall environment: The fully interactive physics lecture. *American Journal of Physics*, *70*(6), 639–654. doi:10.1119/1.1463739

Mendo-Lázaro, S., León-del-Barco, B., Felipe-Castaño, E., Polo-del-Río, M. I., & Iglesias-Gallego, D. (2018). Cooperative team learning and the development of social skills in higher education: The variables involved. *Frontiers in Psychology*, *9*. PMID:30186208

Mendro, R. L. (1998). Student achievement and school and teacher accountability. *Journal of Personnel Evaluation in Education, 12*, 257–267.

Meschitti, V. (2018). Can peer learning support doctoral education? Evidence from an ethnography of a research team. *Studies in Higher Education*, 1–13.

Meyer, J., & Land, R. (2003). *Threshold concepts and troublesome knowledge: Linkages to ways of thinking and practicing within the disciplines*. Edinburgh, UK: University of Edinburgh.

Miller, G., Sharma, S., Donald, C., & Amor, R. (2013). Developing a building information modelling educational framework for the tertiary sector in New Zealand. In *IFIP International Conference on Product Lifecycle Management* (pp. 606-618). Springer. 10.1007/978-3-642-41501-2_60

Moore, P., & Hampton, G. (2015). 'It's a bit of a generalisation, but …': Participant perspectives on intercultural group assessment in higher education. *Assessment & Evaluation in Higher Education*, *40*(3), 390–406. doi:10.1080/02602938.2014.919437

Moss, K., & Crowley, M. (2011). Effective learning in science: The use of personal response systems with a wide range of audiences. *Computers & Education*, *56*(1), 36–43. doi:10.1016/j.compedu.2010.03.021

Müller, C., Waha, K., Bondeau, A., & Heinke, J. (2014). Hotspots of climate change impacts in sub-Saharan Africa and implications for adaptation and development. *Global Change Biology*, *20*(8), 2505–2517. doi:10.1111/gcb.12586 PMID:24796720

Nasirian, A., Arashpour, M., & Abbasi, B. (2018). Multiskilled human resource problem in off-site construction. In *35th International Symposium on Automation and Robotics in Construction and International AEC/FM Hackathon: The Future of Building Things, ISARC 2018*. International Association for Automation and Robotics in Construction I.A.A.R.C).

Nawari, N. O. (2010). Intelligent Design in AEC Education. *ITcon, 15*, 306–317.

Ness, D., Kim, K., Swift, J., Jenkins, A., Xing, K., & Roach, N. (2019). Cradle to cradle building components via the cloud: a case study. In *6th CIB International Conference on Smart and Sustainable Built Environments - SASBE2018*. Emerald Publishing Group.

New Jersey Engineers Australia. (2005). *Getting it Right First Time: A Plan to Reverse declining standards in project design documentation within the building and construction industry*. Engineers Australia. Available at: http://codebim.com/wp-content/uploads/2013/06/Getting-It-Right-The-First-Time.pdf

Newton, S., Hardie, M., Lowe, R., Pickersgill, S., Rameezdeen, R., Srivastava, A., & Zillante, G. (2015). Situational eLearning: the new potential for digital technology. In RICS COBRA AUBEA 2015: The Construction, Building and Real Estate Research Conference of the Royal Institution of Chartered Surveyors. The Australasian Universities' Building Educators Association Conference.

Ng, W., & Hanewald, R. (2009). Concept maps as a tool for promoting online collaborative learning in virtual teams with pre-service teachers. In Handbook of Research on Collaborative Learning Using Concept Mapping. IGI Global.

Nicol, D. J., & Macfarlane-Dick, D. (2006). Formative assessment and self-regulated learning: A model and seven principles of good feedback practice. *Studies in Higher Education, 31*(2), 199–218. doi:10.1080/03075070600572090

Niculae, R., & Duda, M. (2015). Gamification as a Means of Getting from e-Learning to Experience Learning in Architecture. Carol I' National Defence University.

Noguera, I., Guerrero-Roldán, A. E., & Masó, R. (2018). Collaborative agile learning in online environments: Strategies for improving team regulation and project management. *Computers & Education, 116*, 110–129. doi:10.1016/j.compedu.2017.09.008

O'Neill, G., & McMahon, T. (2005). *Student-centred learning: What does it mean for students and lecturers*. Academic Press.

Odeh, S., McKenna, S., & Abu-Mulaweh, H. (2017). A unified first-year engineering design-based learning course. *International Journal of Mechanical Engineering Education, 45*(1), 47–58. doi:10.1177/0306419016674508

Oja, S., & Smulyan, L. (1989). *Collaborative action research: a developmental approach*. Falmer Press.

Pant, I., & Baroudi, B. (2008). Project management education: The human skills imperative. *International Journal of Project Management, 26*(2), 124–128. doi:10.1016/j.ijproman.2007.05.010

Park, K. S., & Kim, K. P. (2014). Essential BIM input data study for housing refurbishment: Homeowners' preferences in the UK. *Buildings, 4*(3), 467–487. doi:10.3390/buildings4030467

Peacock, A. D., Jenkins, D. P., & Kane, D. (2010). Investigating the potential of overheating in UK dwellings as a consequence of extant climate change. *Energy Policy, 38*(7), 3277–3288. doi:10.1016/j.enpol.2010.01.021

Peel, D. (2010) Editorial: Making an Impact in the Built Environment? *Journal for Education in the Built Environment, 5*(1), 1-6.

Pegg, A., Waldock, J., Hendy-Isaac, S., & Lawton, R. (2012). *Pedagogy for employability. Pedagogy For Employability*. York, UK: Higher Education Academy.

Penttila & Elger. (2006). New professional profiles for international collaboration in design and construction. *26th eCAADe Conference Proceedings.*

Piekkari, R., Welch, C., & Paavilainen, E. (2009). The Case Study as Disciplinary Convention. *Organizational Research Methods, 12*(3), 567–589. doi:10.1177/1094428108319905

Poelmans, S., & Wessa, P. (2015). A constructivist approach in a blended e-learning environment for statistics. *Interactive Learning Environments, 23*(3), 385–401. doi:10.1080/10494820.2013.766890

Porritt, S., & Cropper, P., Shao, L., & Goodier, C. (2013). Heat wave adaptations for UK dwellings and development of a retrofit toolkit. *International Journal of Disaster Resilience in the Built Environment, 4*(3), 269–286. doi:10.1108/IJDRBE-08-2012-0026

Pressman, A. (2007). Integrated Practice in Perspective: A New Model for the Architectural Profession. *Architectural Record.* Retrieved from http://archrecord.construction.com/practice/projDelivery/0705proj-3.asp

Productivity, C. (2015). *Teicholz*. Retrieved from https://projectproduction.org/presentations/2015/2015-PPI-Teicholz.pdf

Puolitaival, T., & Forsythe, P. (2016). Practical challenges of BIM education. *Structural Survey, 34*(4/5), 351–366. doi:10.1108/SS-12-2015-0053

Pyecha, J. (1988). *A case study of the application of non-categorical special education in two states*. Chapel Hill, NC: Research Triangle Institute.

Radel, J. (1999). *Creating Effective Posters*. Retrieved 1 June, 2011, from www.kumc.edu/SAH/OTEd/jradel/Poster_Presentations/PstrStart.html

Rae, D., Gee, S., & Moon, R. (2009). Creating an Enterprise Culture in a University: The Role of an Entrepreneurial Learning Team. *Industry and Higher Education, 23*(3), 183–197. doi:10.5367/000000009788640279

Ramaprasad, A. (1983). On the definition of feedback. *Behavioral Science, 28*(1), 4–13. doi:10.1002/bs.3830280103

Ramsden, P. (1992). *Learning to Teach in Higher Education*. Routledge.

Raufflet, E. (2013). Integrating sustainability in management education. *Humanities (Washington)*, *2*(4), 439–448. doi:10.3390/h2040439

Reimann, N. (n.d.). *Getting to grips using posters for formative feedback.* Northumbria University. Retrieved from http://www.northumbria.ac.uk/static/5007/cetlpdf/signpost9.pdf

Renolds & Simmonds. (1984). *Presentation of data in Science, Publications slides, posters, overhead projections, tape=slides, television, Principles and practices for authors and teachers.* Dordrecht, The Netherlands: Marinus Nijjhoff Publishers.

Ren, Z., Wang, X., & Chen, D. (2014). Heat stress within energy efficient dwellings in Australia. *Architectural Science Review*, *57*(3), 227–236. doi:10.1080/00038628.2014.903568

RICS SBE. (2014). *RICS SBE.* Retrieved August 20, 2018, from http://ricssbe.org/about-us/about-rics-sbe/overview/

Rittel, H. W. J., & Webber, M. M. (1972). *Dilemmas in a General Theory of Planning.* Working Paper No. 72 – 194. University of California, Berkeley.

Rogers, C. (1965). *Client-centred Therapy.* Houghton-Mifflin.

Romer, K., & Whipple, W. (1990). Collaboration across the Power Line. *College Teaching*, *39*(2), 66–70. doi:10.1080/87567555.1991.9925490

Rooney, K. (2017). *BIM education - global. 2017 update report.* ICIS Integrated Construction. Retrieved May 28, 2017, from http://www.canbim.com/single-post/2017/05/29/NATSPEC-Global-BIM-Education-Report-2017

Rooney, K. (2018). *NATSPEC Construction Information; BIM Education Global 2018 Update Report.* Retrieve from https://www.icis.org/wp-content/uploads/2018/05/BIM-Education-Global-2018-Update-Report-V5.0.pdf

Rowe, A.D., & Zegwaard, K.E. (2017). *Developing graduate employability skills and attributes: Curriculum enhancement through work-integrated learning.* Academic Press.

Sabongi, F. J. (2009). *The integration of BIM in the undergraduate curriculum: An analysis of undergraduate courses.* Paper presented at the International Proceedings of the 45th ASC (Associated Schools of Construction) Annual Conference, Gainesville, FL.

Sacks, R., & Pikas, E. (2013). Building information modeling education for construction engineering and management. I: Industry requirements, state of the art, and gap analysis. *Journal of Construction Engineering and Management*, *139*(11), 04013016. doi:10.1061/(ASCE)CO.1943-7862.0000759

Sadler, D. R. (1989). Formative assessment and the design of instructional systems. *Instructional Science*, *18*(2), 119–144. doi:10.1007/BF00117714

Saman, W., Boland, J., Pullen, S., de Dear, R. J., Soebarto, V., Miller, W. F., ... Pockett, J. (2013). *A framework for adaptation of Australian households to heat waves.* National Climate Change Adaptation Research Facility.

Sandberg, K. E. (1995). Affective and cognitive features of collaborative learning. Review of Research and Developmental Education, 6(4).

Santamouris, M. (2014). On the energy impact of urban heat island and global warming on buildings. *Energy and Building*, *82*, 100–113. doi:10.1016/j.enbuild.2014.07.022

Savage, S., Davis, R., & Miller, E. (2010). *Professional Education in Built Environment and Design*. Australian Learning and Teaching Council.

Sawyer, S., & Tapia, A. (2005). The sociotechnical nature of mobile computing work: Evidence from a study of policing in the United States. *International Journal of Technology and Human Interaction*, *1*(3), 1–14. doi:10.4018/jthi.2005070101

Scager, K., Boonstra, J., Peeters, T., Vulperhorst, J., & Wiegant, F. (2016). Collaborative learning in higher education: Evoking positive interdependence. *CBE Life Sciences Education*, *15*(4), ar69. doi:10.1187/cbe.16-07-0219 PMID:27909019

Schaber, P., McGee, C., & Jones, T. (2015). Building student community in a hybrid program. *Occupational Therapy in Health Care*, *29*(2), 102–114. doi:10.3109/07380577.2015.1017134 PMID:25821882

Scharle, A., & Szabó, A. (2000). *Learner Autonomy: A guide to developing learner responsibility*. Cambridge University Press.

Schon, D. A. (1983). *The Reflective Practitioner: how professionals think in action*. New York: Basic Books.

Scott, J. L., Mistry, S., Moxham, B. J., & Rutherford, S. M. (2014). Using web 2.0 technology to support and enhance collaborative activity outside of the taught curriculum in higher education. In Collaborative Learning: Theory, Strategies and Educational Benefits. Nova Science Publishers, Inc.

Scott, L. (2016). Embedding Action Research in the Built Environment: An Action Approach. In Research Methodology in the Built Environment. Routledge.

Scott, L., & Fortune, C. (2010). An Investigation of Assessment Practices in Built Environment Education – the Views of Senior Academics. *Proceedings Associated Schools of Construction 46th Annual International Conference*.

Scott, L., & Fortune, C. (2011). Reshaping assessment in Built Environment undergraduate programs - the challenge. *Proceedings Associated Schools of Construction 47th Annual International Conference*.

Scott, L., & Fortune, C. (2012). Towards Assessment Practices 'for' Learning in Irish Built Environment Higher Education. *Journal for Education in the Built Environment*.

Scott, L. (2015). The Changing Landscape of Construction Higher Education. *International Journal of Construction Education and Research*, *11*(2), 78–78. doi:10.1080/15578771.2015.1022454

Scott, L. M. (2016). Theory and research in construction education: The case for pragmatism. *Construction Management and Economics, 34*(7-8), 552–560. doi:10.1080/01446193.2016.1151539

Scott, L., & Fortune, C. (2009). Promoting student centered learning: the use of portfolios in the assessment of learning on an undergraduate construction management program. *Proceedings of the Associated Schools of Construction Conference.*

Scott, L., & Fortune, C. (2013). Towards the Improvement of the Student Experience of Assessment and Feedback in Construction Management Education. *European Journal of Engineering Education, 38*(6), 661–670. doi:10.1080/03043797.2013.766675

Scott, L., Harvey, J., & Hayes, N. (2007). *Enhancing Student Learning through Assessment and Feedback A Toolkit Approach.* Dublin Institute of Technology.

Scriven, M. (1967). The methodology of evaluation. In *Perspectives on Curriculum Evaluation (AERA Monograph Series – Curriculum Evaluation).* Chicago: Rand McNally and Co.

Serçe, F. C., Nur Alpaslan, F., Swigger, K., Brazile, R., Dafoulas, G., & Lopez, V. (2010). Strategies and guidelines for building effective distributed learning teams in higher education. *2010 9th International Conference on Information Technology Based Higher Education and Training.*

Shelbourn, M., Macdonald, J., McCuen, T., & Lee, S. (2017). Students' perceptions of BIM education in the higher education sector: A UK and US perspective. *Industry and Higher Education, 31*(5), 293–304. doi:10.1177/0950422217725962

Shelbourn, M., Macdonald, J., & Mills, J. E. (2016). Developing an international framework for BIM education in the HE sector. *Proceedings of the 10th Academic Interoperability Coalition (AIC) BIM Symposium.*

Shuell, T. (1986). Cognitive conceptions of learning. *Review of Research in Education, 19*, 405–450.

Simon, H. A. (1984). The Structure of Ill-structured Problems. In N. Cross (Ed.), *Developments inDesign Methodology* (pp. 317–327). Chichester, UK: J. Wiley & Sons.

Simpson, B. J. (1966). The classification of educational objectives: Psychomotor domain. *Illinois Journal of Home Economics, 10*(4), 110–144.

Singh, S. R. S. (2018). *Sustainability in Construction Management Education: An Indian Perspective.* Cham: Springer. doi:10.1007/978-3-319-73293-0_9

Skelley, J. W. (2014). A small group assignment gives students a novel opportunity to demonstrate current clinical controversies in a self-care course. *American Journal of Pharmaceutical Education, 78*(10), 1–8. doi:10.5688/ajpe7810193 PMID:25657380

Smith, M. A., & Kellogg, D. L. (2015). Required Collaborative Work in Online Courses: A Predictive Modeling Approach. *Decision Sciences Journal of Innovative Education, 13*(4), 563–581. doi:10.1111/dsji.12078

Stache, J. M., Barry, B. E., Klosky, J. L., & Freisinger, G. M. (2017). Assessing individual temperament and group performance in a project-based learning experience. In *124th ASEE Annual Conference and Exposition*. American Society for Engineering Education. 10.18260/1-2--27621

Starzyk, G. F., & McDonald, M. (2010). The Collaborative Dance: Only Three Steps. *Proc of the BIM-Related Academic Workshop.*

Stevens, M. (2012). *The Construction MBA*. McGraw-Hill Education.

Streiner, S., Besterfield-Sacre, M., & Donovan, S. (2016). Using simulation and structured group work to address statistical misconceptions. In *46th Annual Frontiers in Education Conference, FIE 2016*. Institute of Electrical and Electronics Engineers Inc.

Stronge, J. H., & Tucker, P. D. (2000). *Teacher evaluation and student achievement*. Washington, DC: National Education Association.

Surentheran, T. (2014). Use of an Electronic Voting System to facilitate interactive engagement and enhance active learning of undergraduate students in Biomedical Science. *Investigations in University Teaching and Learning, 9*, 111-121.

Suwal, S., & Singh, V. (2018). Assessing students' sentiments towards the use of a Building Information Modelling (BIM) learning platform in a construction project management course. *European Journal of Engineering Education*, *43*(4), 492–506. doi:10.1080/03043797.2017.1287667

Tangney, S. (2014). Student-centred learning: A humanist perspective. *Teaching in Higher Education*, *19*(3), 266–275. doi:10.1080/13562517.2013.860099

Taras, M. (2002). Using assessment for learning and learning from assessment. *Assessment & Evaluation in Higher Education*, *27*(6), 501–510. doi:10.1080/0260293022000020273

Taras, M. (2005). Assessment–summative and formative–some theoretical reflections. *British Journal of Educational Studies*, *53*(4), 466–478. doi:10.1111/j.1467-8527.2005.00307.x

Teicholz. (2015). *Construction Productivity Index Graph*. Indexes of labor productivity from construction and non-farm industries, 1964 -2004, indicates no productivity gain in the construction industry.

Thrall, G. C., Coverdale, J. H., Benjamin, S., Wiggins, A., Lane, C. J., & Pato, M. T. (2016). A Randomized Controlled Trial of Team-Based Learning Versus Lectures with Break-Out Groups on Knowledge Retention. *Academic Psychiatry*, *40*(5), 755–760. doi:10.100740596-016-0501-7 PMID:26883529

Tinker, A. A. (2003, April). "Greening" the construction curriculum. *ASC Proceedings of the 39th Annual Conference,* 113-118.

Topping, K., & Stewart, E. (1998). *Peer Assisted Learning*. Lawrence Erlbaum Associates.

Tran, D., Nguyen, L., & Faught, A. (2017). Examination of communication processes in design-build project delivery in building construction. *Engineering, Construction, and Architectural Management, 24*(6), 1319–1336. doi:10.1108/ECAM-12-2015-0192

Tucker, R., Abbasi, N., Thorpe, G., Ostwald, M., Williams, A., Wallis, L., & Kashuk, S. (2014). *Enhancing and assessing group and team learning in architecture and related design contexts.* Office for Learning and Teaching.

UN Environment and International Energy Agency. (2017). *Towards a zero-emission, efficient, and resilient buildings and.* UNEP.

University of Melbourne. (2018). *Academic Posters, All about academic posters.* Retrieved 03/07/2018 from https://services.unimelb.edu.au/__data/assets/pdf_file/0007/470059/Academic_posters_Update_051112.pdf

University of Wisconsin. (n.d.). *The Writing Cneter Academic and Professional Writing: Creating Poster Presentations.* Retrieved from https://writing.wisc.edu/Handbook/PosterPresentations.html

van Gelder, T. (2009). *What is Decision Mapping?* Retrieved February 1, 2017, from https://timvangelder.com/2009/09/08/what_is_decision_mapping/

Van Leeuwen, A. (2016). Learning analytics in a flipped University course. In *6th International Conference on Learning Analytics and Knowledge, LAK 2016.* Association for Computing Machinery. 10.1145/2883851.2883874

Vanderbilt University. (2018). *Centre for Teaching.* Retrieved from CFT.vanderbilt.edu: https://cft.vanderbilt.edu/2010/08/new-cft-guide-on-sustainability-and-pedagogy/

Voelkel, S., & Bennett, D. (2014). New uses for a familiar technology: Introducing mobile phone polling in large classes. *Innovations in Education and Teaching International, 51*(1), 46–58. doi:10.1080/14703297.2013.770267

Von Glaserfeld, E. (1995). A constructivist approach to teaching. In L. Steffe & J. Gale (Eds.), *Constructivism in education* (pp. 3–16). Lawrence Erlbaum Associates.

Vygotsky, L. (1978). *The Mind in Society – The Development of Higher Psychological Processes.* The Harvard University Press.

Vygotsky, L. S. (1978). *Mind in society: the development of higher psychological processes.* Harvard University Press.

Wang, E. L., & Vollstedt, A. M. (2014). A method for adjusting group-based grades. In *121st ASEE Annual Conference and Exposition: 360 Degrees of Engineering Education.* Indianapolis, IN: American Society for Engineering Education.

Wang, J., Tsiligiris, V., & Hartley, R. (2018). Perceptions of the benefits of the A3 planner in facilitating project-based learning in accounting education. *Accounting Education.*

William, D. (2000b). *Integrating summative and formative functions of assessment.* Keynote address to the European Association for Educational Assessment, Prague, Czech Republic.

Williams, A., Sher, W., Simmons, C., Dosen, A., & Pitt, B. (2009). *Construction Education in Australia: a review of learning and teaching challenges and opportunities.* Australian Learning and Teaching Council.

Wilson, K., & Fowler, J. (2005). Assessing the impact of learning environments on students' approaches to learning: Comparing conventional and action learning designs. *Assessment & Evaluation in Higher Education, 30*(1), 87–101. doi:10.1080/0260293042003251770

Wlodkowski, R. J. (1985). *Enhancing Motivation to Learn.* San Francisco: Josey- Bassey.

Wolter, B. H., Lundeberg, M. A., Kang, H., & Herreid, C. F. (2011). Students' perceptions of using personal response systems ("clickers") with cases in science. *Journal of College Science Teaching, 40,* 14–19.

Wong, N. H., & Chen, Y. (2008). *Tropical urban heat islands: climate, buildings and greenery.* Routledge. doi:10.4324/9780203931295

Wood, D. (1988). *How children think and dream.* Oxford, UK: Basis Blackwell.

Wood, H. A., & Middleton, D. J. (1976). An experimental evaluation of four face-to-face teaching strategies. *International Journal of Behavioral Development, 1*(2), 131–147. doi:10.1177/016502547800100203

Woolard, N. A. (2018). Rethinking management group projects with the adaptive leadership model: The lesson is the process. *Journal of Education for Business, 93*(8), 392–402. doi:10.1080/08832323.2018.1496895

World Commission on Environment and Development. (1987). *Our Common Future.* UN.

Wright, L., & Jones, A. (2007). *Good Practice and Innovation in the Assessment of Students in Built Environment Subject in Higher Education.* Liverpool John Moores University.

Yalcinkaya, M., & Singh, V. (2015). Patterns and trends in building information modeling (BIM) research: A latent semantic analysis. *Automation in Construction, 59,* 68–80. doi:10.1016/j.autcon.2015.07.012

Yannibelli, V., & Amandi, A. (2017). A hybrid evolutionary algorithm based on adaptive mutation and crossover for collaborative learning team formation in higher education. *18th International Conference on Intelligent Data Engineering and Automated Learning.*

Yorke, M. (2006). *Employability in higher education: what it is-what it is not* (Vol. 1). York, UK: Higher Education Academy.

Yuretich, R. F., & Kanner, L. C. (2015). Examining the effectiveness of team-based learning (TBL) in different classroom settings. *Journal of Geoscience Education, 63*(2), 147–156. doi:10.5408/13-109.1

Zarei, H. (2018). *Participants' power asymmetry in public infrastructure projects* (PhD thesis). University of Melbourne.

Zheng, L. (2015). A systematic literature review of design-based research from 2004 to 2013. *Journal of Computer Education, 2*, 399. doi:10.100740692-015-0036-z

Zitzelsberger, H., Campbell, K. A., Service, D., & Sanchez, O. (2015). Using wikis to stimulate collaborative learning in two online health sciences courses. *The Journal of Nursing Education, 54*(6), 352–355. doi:10.3928/01484834-20150515-08 PMID:26057431

Zschocke, K., Wosnitza, M., & Bürger, K. (2016). Emotions in group work: Insights from an appraisal-oriented perspective. *European Journal of Psychology of Education, 31*(3), 359–384. doi:10.100710212-015-0278-1

Zuo, J., Pullen, S., Palmer, J., Bennetts, H., Chileshe, N., & Ma, T. (2015). Impacts of heat waves and corresponding measures: A review. *Journal of Cleaner Production, 92*, 1–12. doi:10.1016/j.jclepro.2014.12.078

About the Contributors

Sherif Mostafa is a Lecturer in Construction and Engineering Project Management at Griffith University. He is leading, developing, coordinating, and lecturing construction and project management subjects. Sherif areas of expertise including Lean-Six-Sigma and BIM integration within prefabricated buildings, advanced technologies in offsite construction and dynamic simulation of engineering systems.

* * *

Mehrdad Arashpour is a Senior Lecturer and Director of the Master of Infrastructure Engineering and Management Program at Monash University. He is undertaking research on prefabricated structures and processes, Building Information Modelling (BIM), and automation and optimisation of design, manufacturing and assembly.

Anurita Bhatnagar has experience in with NGO working with sustainability projects and currently she is an assistant professor at School of Construction, RICS SBE.

Jennifer E. Day, Ph.D., is a Senior Lecturer at the University of Melbourne, Australia. Her areas of expertise include civil engineering and economics. Day received her Doctorate from the University of California, Berkley and has authored or co-authored publications ranging in topic from "Giga" projects to productivity to dispute resolution. She has presented her research at dozens of conferences in front of other experts throughout the world. Her engagements include the World Bank and national governments. Day's advising work focuses on Organizational Development, Strategic Planning, Business Evaluations, and Organizational Productivity.

Colin Duffield is Professor in Engineering Project Management and Deputy Head of the Department of Infrastructure Engineering at The University of Melbourne. He is also a fellow of the Law School (where he delivers a master level unit in infrastructure procurement) and a Director of Infrastructure Australia. Colin has extensive experience in the governance of long term contracts and the interaction between policy, technical matters, risk management, financing and contractual arrangements as they apply to infrastructure. This expertise has been gained through: an involvement in infrastructure delivery for public and private clients (both as a practitioner and researcher); and as an adviser to projects on risk and project structuring; and as an independent reviewer and researcher of major engineering contracts.

Obuks Ejohwomu is the Director Commercial Project Management, Admissions Tutor, Management of Projects (MSc) Programmes and Research Group PGR Lead in the School of Mechanical, Aerospace and Civil Engineering. He is also the Unit Lead for Commercial and Procurement. Prior to joining MACE he has worked as a contracts manager on several building construction projects totalling over a N100m (Lukayson Ventures Ltd), systems engineer in the telemetry and laboratory business area of Scottish Water, Glasgow, Research Assistant on a Black Country learning and skills action research project (University of Wolverhampton), Research Fellow on the EPSRC Health and Care Infrastructure Research and Innovation Centre (HaCIRIC), Programme Manager (W D Consults Ltd), Lecturer in Civil Engineering (Covenant University), Senior Lecturer in Project Management and Head, Department of Management Technology (Bells University of Technology). In the past he has consulted for the likes of Coffey International on DFID projects (Growth and Employment in States (GEMS)) in Nigeria, National Petroleum Investment Management Services (NAPIMS), Nigerian National Petroleum Corporation Staff Cooperative Multipurpose Society, and Pipelines and Products Marketing Company Limited (PPMC). He obtained his BEng degree in Civil Engineering from UNIBEN, MSc in Systems Engineering from Heriot-Watt, and PhD in Construction Engineering and Management from Wolverhampton. His research focus is in effective communication, procurement, social value diffusion, diversity, productivity and competitive advantage. He has served as external PhD examiner to KNUST, Ghana, University of Lincoln, University of Wolverhampton, University of Bedfordshire, Nelson Mandela University and University of West of England. He is a Carnegie and APM grant winner. He is also a grant reviewer for Newton Funds, Carnegie Africa Diaspora Fellowship programme and National Research Foundation in South Africa. When the opportunity is available I enjoy networking and walking.

Parisa Farzanehfar is an academic staff in University of Melbourne. Her research interest is focused on higher education research.

Susan George is a qualified Architect has good number experience in architectural consultancy and worked on many sustainability projects. She currently woks as adjunct associate professor at School of Construction, RICS school of Built Environment.

Jeremy Gibberd has over 20 years of experience working in a range of a range of roles within the planning, design and management of the built environment. He has developed specialist expertise in sustainable built environments, education and community architecture, building performance analysis and facilities management. He has a wide knowledge of relevant legislation, technological developments and best practice in these areas. As a visiting professor, Jeremy lectured on environmental design, advanced technology and design at SCAD in the USA and has developed and taught courses on urban design, architectural design, materials, sustainability, energy and facilities management with organisations such as GIZ, ICLEI, the Green Building Council of South Africa (GBCSA), the South African Property Owners Association (SAPOA) the University of the Pretoria and the University of the Free State. He has examined at masters and doctorate levels at Queensland University of Technology (Australia), University of Cape Town and University of Johannesburg (South Africa). He regularly works as a research scientist and consultant to government, business and community organisations. Dr Gibberd has published widely and presented as invited or keynote speaker at numerous international meetings and conferences. More information on his work is available at jeremygibberd.com

Nicky Harris is a Quantity Surveyor by profession with 15 years industrial experience working on a range of large commercial developments such as schools, hospital, pharmaceutical laboratories, infrastructure works and sustainable communities. As part of her role at the University of Salford she is responsible for encouraging all students to engage in some form of industrial practice, whether it be a summer internship, work shadowing opportunity or a yearlong industrial placement with an industrial partner. Nicky is very committed to active learning and co-ordinates site visits for all students across the school to afford them the opportunity to gain an insight in the working practice and technologies. She teaches the following courses: Multi-disciplinary project, Professional Practice, An introduction to Construction Management and Construction Technology.

Jennifer Harvey has degrees in Architecture, Classics and Education and has taught and researched widely in the fields of Design Studio and Architectural Theory. She has particular expertise in the relationship between creative practice and educational pedagogy.

Felix Kin Peng Hui is a Senior Lecturer in Engineering Management and an Academic Specialist in the Department of Infrastructure Engineering. He is also the Academic Coordinator for the Master of Engineering Management program at the University of Melbourne. Dr Hui.is an experienced engineer with over 25 years of industry experience in managing continuous improvement projects, factories, and business units. He has held senior management positions in a range of manufacturing industries which includes machine tools, precision engineering and semiconductors. He has also consulted to organisations seeking continuous improvements to optimise their operational efficiency. His research interests are in the areas of operational process optimisation, operational efficiency, lean systems, organisational development and team dynamics in engineering projects. He is currently a lead investigator on mega-infrastructure projects research and on prefabricated modular construction research. Dr Hui is a registered Professional Engineer (Singapore), an ASEAN Chartered Professor Engineer, a Fellow of Institute of Managers and Leaders (formerly The Australian Institute of Management) and a Council member of Lean Construction Institute of Australasia (Victoria).

Anupam Jain is an Architect by profession and contributed much to the topic of sustainability, a member of green building and he is adjunct visiting faculty at School of Construction at RICS school of Built Environment.

Ki Pyung Kim, PhD, PMP, PMI-SP, MCTS, is a Lecturer in the Construction Project Management program at the University of South Australia. He obtained his PhD from the Aston University in 2014 for research on integrating Building Information Modelling (BIM) into housing refurbishment. Prior to his PhD studies, he taught project management courses and relevant tools including MS Project and Primavera 6 from 2008 to 2011, and worked as an Information System manager at University of Alaska Anchorage. His research interests and expertise lie in BIM system and application, sustainable housing refurbishment, data/information management and construction time management. He was awarded 2018 Division of ITEE Early Career Teacher Award; Highly Commended Academic Paper in BIM for QS at the Pacific Association of Quantity Surveyors Congress 2016; Excellence in Volunteer Recognition for PMI Knowledge Asset Localization in South Korea 2015; and the Best Mention Prize at the BIM competition 2013 hosted by buildingSMART Singapore. Recently, he was awarded the Augmented Reality Visualisation of BIM for Improved Productivity of Construction funded by the University of South Australia; DFTA Australia-Korea Foundation Research Funding in 2018 and 2017; China-Australia Centre for Sustainable Urban Development funding for Waste Management using BIM in 2017; University of South Australia Seed Funding for Big Data Analytics for Waste Management using BIM in 2017; and ARUP Global Research Challenge Grant 2017 for Cloud-BIM Platform for Reusable Building Components.

Julia Lamborn received her BEng (Civil) GradDipChemEng; and MEng and PhD in Environmental Engineering, all from Swinburne University of Technology. Her PhD thesis was on landfill gas modelling. Julia is a civil, chemical and environmental engineer. She was employed by the State Electricity Commission of Victoria for 10 years as the Cooling Tower Thermal Design Engineer. During that time she designed the Loy Yang B natural draft cooling towers, supervised their construction, undertook the commissioning and oversaw the operation of the towers once the power station went into full production. From 1990 – 2015, she has been employed at Swinburne University of Technology in the areas of civil, chemical and environmental engineering. She held multiple leadership roles during her time there including Director, Engineering Accreditation, Associate Dean (Education) and Associate Dean (Learning Innovation). In 2015 she was appointed as Professor in the Department of Civil Engineering and Associate Dean (Education) for the Faculty of Engineering at Monash University. She has been very active within Engineers Australia and on State Government Statutory Authorities and Panels for over 30 years.

Kenneth Park is currently Deputy Head of Engineering, Systems & Management at Aston University, UK and has over 24 years construction research and industrial experience internationally. His research and expertise lie in the areas of construction management, whole-life management, BIM application, and process improvement. He is currently developing housing information modelling based on whole-life performance via Scan-to-BIM to improve performance and reduce environmental and financial impact. He is a Chartered Construction Manager and Chartered Surveyor in the UK.

Sean Pickersgill teaches within the Architecture Program in the School of Art Architecture and Design at the University of South Australia. he has published widely on the intersection of digital culture and the architecture and construction industry.

Rameez Rameezdeen is Program Director for Construction Management and Economics programs with the School of Natural and Built Environments, UniSA. He coordinates the design and delivery of many courses of study on both undergraduate and postgraduate programs and has responsibility for the teaching of quantity surveying courses within the above degree programs.

Lloyd Scott is Academic Advisor and Partnership Co-Ordinator in the School of Surveying and Construction Management at Technological University Dublin (TUD). He joined the TUD (formerly DIT) as a lecturer in Construction Managementand Technology in 2000. He is currently supervising 5 PhD researchers. Apart from his lecturing, supervision, research and academic administrative duties, Lloyd has completed a PhD in the field of Built Environment Education and has developed a framework for assessment led learning strategies for Built Environment education. In 2014 he accepted the position of 'Professor of Practice' at the Haskell and Irene Lemon Construction Science Division in the College of Architecture at Universityof Oklahoma. Along with this he has produced many peer reviewed conference and journal papers. He serves as the Associate Editor of the International Journal of Construction Education and Research. His research interests include modern approaches to thermal performance in domestic construction, development of sustainable energy sources and their practical application in Ireland and project delivery methods for a sustainable environment in Ireland. He also serves on the editorial board of International Journal of Building Pathology and Adaptation an academic journal that publishes contemporary and original research in building pathology and building forensics, refurbishment and adaptation. In 2014, Lloyd accepted the position of Research Fellow at the Sustainability Institute at Leeds Beckett University.

Mark Shelbourn is Associate Professor in Building Information Modelling (BIM) & Digital Technologies at the University of Salford and has been researching and teaching in the area of BIM, collaborative working and digital technologies for over 10 years. His specific interest are people's behaviours that enable them to better collaborate as individuals and projects. Mark holds a PhD in Computer-Aided-Learning in Building Pathology from the University of Salford (Sept 2002. He also studied an MSc in Information Technology in Property and Construction, also from the University of Salford (July 1998). His first degree was an BSc Honours in Building Surveying at Sheffield Hallam University. Mark is also an adjunct Senior Research Fellow at the University of South Australia where I continue to support students with their studies, particularly the online postgraduate programmes in Building Surveying & Project Management, PhD supervisions, as well as continuing research interests with former colleagues & industry. Mark is currently teaching on the MSc BIM in Digital Built Environments across the different subjects. He is also responsible for the research methods teaching across the Masters suite in the School. Mark is the lead of the Education & Learning in the Built Environment (ELBE) research group within the School of Built environment at University of Salford. His interest lie in improving the learning experience through collaboration and project based learning of all those entering the built environment area.

Matt Stevens, Ph.D., is a Lecturer in Construction Management at Western Sydney University, Australia. Stevens has been advising since 1994, and his industry experience spans over 40 years. Stevens received his Doctorate from the University of Florida's M.E. Rinker Sr. School of Construction Management. He authored two industry books, Managing a Construction Firm on Just 24 Hours a Day (416 pages) and The Construction MBA (512 pages) published by McGraw-Hill, New York. Jennifer Day and he will publish The Construction Contractor's Digest (1013 pages) in 2019.

Joseph V. Thanikal has over 33 years of experience in academic and research. He has done his PhD in Civil Engineering and Postdoctoral fellowship from National Institute of Agricultural Research, France. He has been at leadership roles as Dean of Kumaraguru College of Technology, Coimbatore and Head of School of Built Environment, Caledonian College of Engineering, Muscat, affiliated to Glasgow Caledonian University, UK. Currently, he is the associate Dean and Director of School of Construction at RICS SChool of Built Environment at Noida, UP.

Hamzeh Zarei was born in the Middle East, Hamzeh received his diploma of mathematics in 1998. He then continued bachelor of science in systems analysis and design followed by a master degree in financial engineering. A few years later in 2010, he got a master degree in project management from the University of Calgary. In 2018 he conferred his PhD in infrastructure investment management from the University of Melbourne. Hamzeh possesses Melbourne teaching certificate and also certificate IV in teaching and assessment. Tutorials and lectures have never got him off the hook. Dr Zarei has always been active in the industry. For over a decade, he played key roles in local and international projects in the LNG, infrastructure, and finance. He performs as project manager, project engineer, consultant, and system manager. One of his research interest is engineering management education. Hamzeh is intrigued to find more effective and efficient methods of teaching for engineers.

Index

R

S

T

U

V

W

www.ingramcontent.com/pod-product-compliance
Lightning Source LLC
LaVergne TN
LVHW081327300926
849161LV00008B/1896